# FIELD MAPPING
# FOR GEOLOGY
# STUDENTS

D1761342

# FIELD MAPPING FOR GEOLOGY STUDENTS

## F. Ahmed

*Department of Geology, University of Khartoum*

## D. C. Almond

*Department of Geology, University of Kuwait*

London
GEORGE ALLEN & UNWIN
Boston          Sydney

**George Allen & Unwin (Publishers) Ltd,**
**40 Museum Street, London WC1A 1LU, UK**

George Allen & Unwin (Publishers) Ltd,
Park Lane, Hemel Hempstead, Herts HP2 4TE, UK

Allen & Unwin Inc.,
9 Winchester Terrace, Winchester, Mass. 01890, USA

George Allen & Unwin Australia Pty Ltd,
8 Napier Street, North Sydney, NSW 2060, Australia


First published in 1983

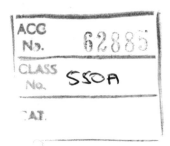
**British Library Cataloguing in Publication Data**

Ahmed, Farouk
　　Field mapping for geology students.
1. Map drawing　　　　　　2. Geology—Maps
I. Title　　　　II. Almond, David C.
551　　GA130
ISBN 0-04-550031-2


**Library of Congress Cataloging in Publication Data**

Ahmed, Farouk.
　　Field mapping for geology students.
Bibliography: p.
1. Geological mapping.　　I. Almond, David C.
(David Coates)　　II. Title.
QE36.A33 1983　　550　　　　　　83-11913
ISBN 0-04-550031-2 (pbk.)


Set in 11 on 13 point Press Roman by
Apsbridge Services Ltd, Nottingham
and printed by Interprint Ltd, Malta

# Preface

The geological map is the most fundamental of all compilations of geological data. Its preparation forms the essential first stage in a wide range of studies, not only those concerned with academic geological problems but also those studies aimed mainly at exploring geophysical, geochemical and hydrogeological features, or with the object of discovering and assessing material resources such as metallic minerals or oil. Moreover, geological maps form basic data for investigations in other fields; surveys of land usage and soils are examples. The ability to prepare such maps is therefore an essential skill for any geologist, whatever his special interests may be. Such work can also give a great deal of pleasure and sense of achievement to the map maker.

The purpose of this book is to explain in simple terms how to approach a geological mapping assignment, from the initial planning stage to the preparation of the final map and accompanying report. We have assumed that the user has a working knowledge of theoretical and laboratory-based geology, and has probably received at least a short introduction to mapping methods in the field. However, we are aware that despite the most careful preparation by his teachers, many a student is worried by his first independent mapping assignment, and we hope that our explicit treatment of basic techniques will help to give him confidence. While our discussion is restricted to elementary techniques, if properly learned these will form a firm base on which to build by experience. Details of more advanced field techniques will be found in the literature listed in the bibliography at the end of this book.

The authors have spent much of their working lives in Africa and the Middle East, and have kept in mind the needs of students learning to map in those regions. It is to such students that we particularly dedicate this work, although there is no reason why students elsewhere should not find our comments useful.

We wish to acknowledge our debt to colleagues at the University of Khartoum and Kingston Polytechnic for numerous discussions of mapping technique. Special thanks are due to John E. Almond, David C. Turner and John B. Wright for careful readings of the manuscript and many useful comments upon it.

F. Ahmed and D. C. Almond
*January 1983*

# Contents

# 1 General procedures

The three major stages of carrying out a mapping project are:

(1)  planning and preparation;
(2)  construction of the map in the field;
(3)  production of the final map and writing the geological report.


## 1.1   The planning stage

Careful planning will make the subsequent field trip more efficient and much easier than would otherwise be the case. Time spent in the field is usually more expensive, more physically tiring and more difficult to arrange than time in the office, so this point is important.

The *purpose* of the project should be clearly understood from the beginning. Is it, perhaps, to make a rapid reconnaissance of the general geology of the area as a guide to further work, or is it to be a more detailed study of part of an area which is already quite well known? Perhaps it is hoped that the project will solve a particular geological problem; or is a much broader coverage of many geological aspects intended? The answers to questions like these will be of great value in making decisions as to the scale on which the mapping will be carried out and the time allotted to complete the fieldwork. A small scale may be chosen for reconnaissance work, perhaps as small as 1 : 250000 (1 cm : 2500 m), whereas more detailed regional work will require a scale of at least 1 : 40000 (1 cm : 400 m) and perhaps as large as 1 : 10000 (1 cm : 100 m). For very detailed mapping in and around active mines, the scales often begin at 1 : 2000 (1 cm : 20 m) and run up to 1 : 200 (1 cm : 2 m) or larger. Many students imagine that small-scale reconnaissance mapping must be easier than detailed work, but this is far from the truth. Indeed, often the best way to tackle a reconnaissance project is to begin by mapping a strip of the country at a moderately large scale (say 1 : 40000), in order to become familiar with the rock types, successions and structures, before using this knowledge to extend the mapping over much larger areas in less detail. The scale of the mapping will, in turn, help to decide the *time* needed to complete the work in the field. This can be a very difficult estimate to make unless one already has experience of a similar type of mapping in a comparable area. Speed of mapping varies greatly, depending on such factors as the experience and ability of the mapper, the degree of rock exposure, the complexity of structure and lithology, ease of access, and the amount of detail needed to satisfy the requirements of the project. In examples drawn from our own experience, detailed mapping of well exposed ground at a scale of $1 : 10000$ progresses at a rate of about $0.5$ km$^2$ per day. Moderately detailed mapping at $1 : 25000$ may cover about $4$ km$^2$ per day, while reconnaissance work at $1 : 250000$ can deal with about $20$ km$^2$ per day. It should be emphasised, however, that these figures form only a very rough guide. Moreover, these are days spent mapping in the field, and additional time must be allowed for travel to and fro with supplies, for vehicle maintenance, and for unforeseen difficulties.

Having decided upon the total duration of the field trip, you can then make arrangements for transport and the purchase of supplies (camp equipment, food and fuel, etc.; see Appendix A). Always work from a carefully prepared list. At this point, it is also necessary to collect together all available maps, aerial photographs, satellite photo prints, and reports which refer to the field area, and make a preliminary assessment of them. We strongly recommend that, before going into the field, you make a preliminary photogeological interpretation and transfer this information onto a general tracing or base map. A good interpretation, showing any visible structural and lithological features, together with outlines of the drainage system and major rock exposures, will save an enormous amount of time and effort in the field. The photo interpretation may also show the best means of access to remote parts of the area, and at this stage it is often a very good idea to talk to somebody who has been in the region before, and question him about the location of any motorable tracks and good drinking-water wells. If aerial photograph cover is not available, it will be necessary to include on the equipment list sufficient topographic surveying instruments to construct a base map. The absence of photograph cover is likely to at least double the field time necessary, and the importance of obtaining aerial photographs where available, and of making good use of them, cannot be over-emphasised.

1 *Select the mapping scale with care, bearing in mind the availability of base maps/photos, and the time to be spent in the field.*
2 *Before going into the field, make a list and assessment of all relevant maps and literature, and prepare a preliminary interpretation of the geology from aerial photographs and satellite prints.*

## 1.2   The map construction stage

When visiting an area for the first time, start by making a brief reconnaissance of the whole block of ground which is to be mapped. This initial inspection may occupy one day or several, depending on the size of the area and the time available, and it will serve to familiarise you with the geographical features, as well as introducing you to some of the rock types and styles of structure which you will be mapping. Beginners mày need to be told not to become discouraged at this stage by the apparent difficulties that the reconnaissance may reveal. Patient work will eventually solve most of these problems, and it is important to understand that even the most experienced field geologist cannot interpret the complexities at first glance. Even at the reconnaissance stage you should be making notes and recording observations on the field maps. Beginners are often reluctant to commit themselves by marking their maps or writing in their notebooks, but they should try to overcome this lack of confidence.

After a day or two spent on the initial reconnaissance, select a suitable stretch of country and start to make detailed observations. To begin with, choose an area which is well exposed and uncomplicated. It is a mistake to tackle the difficult areas at an early stage; they are best left until last, by which time you will be familiar with local rock types and structures. Both the reconnaissance and preliminary aerial photograph interpretation should be of great value in selecting a simple area.

Plan each day's traverse before leaving camp and let somebody know where you are going, at least approximately, as a safety precaution. Some geologists prefer to make their traverses in the form of a loop, mapping all the way round, whereas others like to map only on the outward journey and walk straight back at the end of the day. Your initial plan will often be modified in the course of the day: for instance, during a traverse made across strike to record the succession of rocks you may discover a particularly good 'marker horizon' (i.e. a distinctive bed, useful in tracing successions laterally), and decide to map it along strike for some distance in order to link up with adjacent observations.

As you fill in your map, day by day, make sure that you do not leave any small, unvisited gaps which will require a subsequent special journey. At each exposure make *all* necessary observations and collect *all* the material you may need on the first visit. Do not fall into the habit of leaving some jobs for another day: it should rarely be necessary to revisit exposures unless some problem arises of which you were not aware on your first visit. Although it is wise to cover the ground systematically, it is also necessary to maintain interest if the work is to be done well, and for this reason it is often a good idea to alternate every few days between areas of different lithology and structure.

There are some obvious differences of approach between a reconnaissance project and detailed mapping. When making a detailed map the aim should be to see every exposure, especially in ground which is poorly exposed. In very well exposed ground it may be only necessary to record data on the map at spaced intervals, but one still needs to see every exposure, since the observations should be concentrated at points where there are changes in rock type or structure. A reconnaissance mapping project is often best begun with a period of detailed mapping to gain familiarity with rock types and structural style. Thereafter, traverses will have to be well spaced, often several kilometres apart, and generally selected to cross the strike and visit areas of obvious geological interest. Along the traverse lines themselves, the observations should be in fair detail. Aerial and satellite photographs often provide information which helps in joining up boundaries between traverses and, as a check, it may be possible occasionally to follow a marker horizon on the ground from one traverse line to the next.

Finally, it is important to be aware, while making a geological map, that it differs in a very fundamental way from other kinds of maps, quite apart from the fact that it records different information. For example, a topographical map shows only *observations,* and its quality depends on such properties as accuracy, detail and clarity. A geological map, on the other hand, shows both *observations* and the geologist's *interpretation* of those observations. The symbols which represent rock types and structural measurements made at individual exposures are as much a matter of direct observation as are the hills and valleys on a topographical map, but the way in which geological boundaries and structures are inferred to continue through the intervening unexposed ground is entirely a matter of interpretation. Consequently, no two geologists will produce exactly identical maps if working independently over the same ground, and the quality of a geological map depends not only on accuracy and scope of the observations but also on the intelligence of the interpretation. Interpretation should be carried out almost entirely *in the field,* with all the evidence in front of you, and not left to be done back in the office.

1 *On arrival in the field make a reconnaissance of as much of the area as possible.*
2 *Record your observations from the beginning.*
3 *Start the detailed work in an area where the geology is well exposed and not obviously difficult.*

4 *Plan each day's traverse, and make most of your traverses across the strike rather than along it.*
5 *A geological map should show both observations and interpretations, and distinguish between them. Therefore facts observed at rock exposures are used to infer geological boundaries in the unexposed ground between.*

## 1.3   Map reproduction and the geological report

As the final stage of a mapping project, the geologist is normally required to make a neat copy of his geological map, fully interpreted and illustrated by one or more cross sections, accompanied by an explanation in the form of a geological report. The original field maps and field notebooks are the factual basis from which the final map and report are produced; to this extent they are of more importance than the reproduced material and must be carefully preserved for future reference. The report will also contain additional observations and interpretations which will come from the laboratory examination of material collected in the field. You should be warned that the time necessary to carry out the laboratory work and to prepare the final map and report is often underestimated, so try to make a greater allowance than first thought necessary. It is a good idea to save time by planning, and even writing, some of the geological report at the field-work stage. As already remarked, the interpretation of the field data should certainly be done in the field.

In Chapter 3 of this book you will find detailed advice on how to go about making a neat copy of the map and writing a geological report.

# 2 Field observations and their interpretation

## 2.1 Introduction

This chapter is concerned both with the methods of recording facts observed in the field, and the interpretation of those facts to make geological sense in the form of a map. Although observation and interpretation proceed together, always remember that they are different, and a good geological map will show clearly what is fact and what is interpretation.

One of the main details shown on a geological map are the surface distributions, or outcrops, of the various solid geological formations which occur in the area. More precisely, the map shows where each of these solid formations would be visible at the surface, if all superficial materials (soils, vegetation, sands, gravel, etc.) were to be swept away. Normally, of course, the solid geology will be largely covered by these superficial materials, and the underlying rocks show through only as isolated exposures found, for example, in gullies, cliffs and slabs. The two terms 'outcrop' and exposure' are often confused, and some textbooks use them interchangeably, but we prefer to keep them separate here. One of the main tasks of the geologist is to examine in detail the exposures of solid rock and, from the information they provide, to construct a map which shows the complete outcrops of all the formations present, and the lines of contact between them. In this way the *facts* observed at the exposures are interpreted so as to *infer* the shapes of the entire outcrops.

In this section we discuss first the basic equipment used in fieldwork, including suitable base maps upon which to record and interpret the geological observations. Methods of collecting data from each exposure and plotting them on the base maps are described in section 2.4, while the interpretation of this data to produce an outcrop map is discussed in section 2.5.

## 2.2 Essential equipment

To carry out even the simplest mapping project, the following items are essential:

base map (see section 2.3)
mapping case or boards to support
   the map in the field
compass
clinometer (for measuring angles
   in the vertical plane)

geological hammer
field notebook
metric rule or tape
hand-lens
lead pencils, coloured pencils,
   eraser, protractor, mapping
   pens, drawing ink

Inexpensive mapping baseboards can be made by cutting two sheets of plywood to a size a few centimetres larger all round than the aerial photographs or sections of the base map that will be carried in the field. If standard size aerial photographs are to be used (which measure about 20 cm x 20 cm), it is useful to have sufficient space to use two photographs side by side on the board, so in this case the dimensions might be about 44 cm x 24 cm. The maps or photographs are fastened securely to one of the boards by large rubber bands, while the second board is used to protect the map while travelling between exposures, and is held in place by more rubber bands (Fig. 1b). Suitable bands can be cut from motor tyre inner tubes. If there is likely to be rain, a large transparent plastic bag should be taken into the field to cover the map.

The compass is used mainly for measuring horizontal directions (azimuths) on structures observed at exposures, but it should also be possible to use it to take bearings across country, so it must have either a prismatic or mirror system to allow the sighting of distant points. Liquid-filled prismatic compasses are accurate and easy to use, but are expensive. The Swedish 'Silva' compass and the American 'Brunton' (Fig. 1e and h) use a folding mirror in place of a prism and more practice is required to take good bearings. On the other hand, they are easy to use when taking azimuth readings on structures at the exposure. Both the 'Brunton' and one model of the 'Silva' also contain a needle for measuring angles in the vertical plane. Otherwise, such angles may be measured using a separate clinometer, or the type of carpenter's folding rule which includes a level and a protractor (Fig. 1c).

Geological hammers are made in a number of weights, with either wooden handles or integral steel shafts bound in rubber or leather. Although steel-shafted hammers are expensive, the difficulties which arise from the drying out of wooden handles in hot climates can be avoided. For general purposes, we recommend a weight of about 0.75-1 kg, but lighter hammers can be used on sedimentary rocks.

Field notebooks must withstand hard wear and should have stiff covers and strongly sewn hinges. The notebook should not be too small;

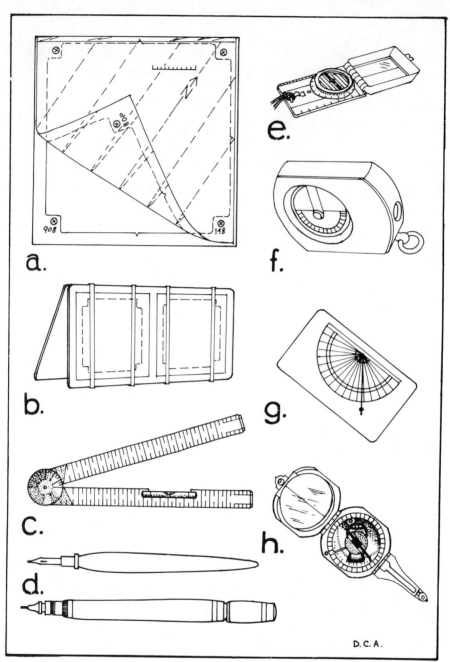

**Figure 1** Some useful items of mapping equipment: (a) an aerial photograph with a transparent overlay; (b) a pair of mapping boards; (c) a carpenter's folding rule with a protractor at the hinge and a level on the lower arm; (d) a simple mapping pen and a Rapidograph pen; (e) a compass made by Silva; (f) an optical clinometer by Suunto; (g) a simple clinometer made from a protractor, a Perspex rectangle and a weighted thread; (h) a compass/clinometer by Brunton.

a size of 20 cm x 12 cm is about right, although some geologists prefer a larger size to accommodate extensive field sketches and sketch maps.

The most useful magnification for a hand-lens is about x10, but some have several elements which can be used singly or in combinations.

Lead pencils, used for rough plotting on the map in the field and for making notes, should be mostly hard (1H to 3H), while a selection of coloured pencils will be needed to shade in the outcrops and exposures as the mapping proceeds. For inking in the day's work at base camp, the best results can be obtained using a simple mapping pen (Fig. 1d), which gives a very fine line but needs some practice, since a light touch is required. The 'Rapidograph' type of pen in the smaller nib sizes may be used for some structural symbols and contact lines but one cannot achieve as fine a line as with a mapping pen.

## 2.3   Base maps

The type of base map upon which the geological information is compiled will be determined partly by the aims of the project (reconnaissance or detailed work, for instance) and partly by what is available: the possibilities range from printed topographical maps, aerial photographs and satellite photo prints, to topographical outline maps surveyed by the geologist himself.

Good topographical maps have a number of advantages - accuracy, clarity and a constant scale, without distortion. Vertical aerial photographs suffer from various distortions due to camera tilt, topographical variation and radial scale change, but they show much more detail than topographical maps, and when viewed stereoscopically they can provide a great deal of useful geological information (see the Bibliography for accounts of photogrammetry and interpretation). Aerial photograph prints are usually to scales of between 1 : 10000 and 1 : 50000, and enlargements can often be made to special order. Satellite photo prints are free from the distortions visible on aerial photographs but are only available at scales of up to 1 : 250000, too small for any but the broadest of reconnaissance surveys, although improvements are promised which may soon allow their use at a scale of 1 : 100000. However, they are often of great use as regional base maps on which can be plotted the main geological features taken from more detailed base maps, as well as the locations of water supplies, inhabited areas and access roads. Moreover, they often reveal large-scale structures that are not obvious from the examination of aerial photographs.

Under ideal circumstances, you will have available an accurate, contoured topographic map as well as stereoscopic coverage of aerial photographs, both at scales equal to or greater than that at which it is intended to produce the final map. However, only a small part of the world is as yet covered by large-scale topographic maps, whereas the coverage by aerial photography and satellite photo prints is very much greater. Aerial photographs alone are more useful than a map alone, since they show more detail and, using them, one can construct a photo-mosaic to cover the whole area, and map either directly onto this or onto the individual prints. Such mosaics are best constructed using a network of properly surveyed fixed points as a control, but even uncontrolled mosaics, made by making the best fit from a number of prints, are better than having no base map at all. In the absence of aerial photographs, satellite photo prints, or accurate topographical maps, you have no alternative but to survey in the topography yourself. For some purposes it is sufficient to use a simple plane table method, or even a system of compass traverses, but for accurate surveys a theodolite and level must be used. When making very large-scale maps around working mines, the mine surveyor's maps can often be adapted for geological use.

If aerial photographs can be obtained, we strongly recommend that the mapping is carried out on transparent overlays securely taped to one edge of each photograph (Fig. 1a). Single weight, matt polyester film is best for this purpose. The information plotted in the field on the individual photograph overlays can be transferred day by day to an overlay of the whole area, prepared by tracing the topography from a photo-mosaic. Before starting to map, each overlay should be marked with the photo-centre and marginal 'fiducial marks' (see Allum 1966 or Lattman & Ray 1965 for the meanings of these and related terms). In this way, the overlay can be kept accurately positioned on the underlying photograph. The complete number of the photograph (i.e. both run and print numbers) should also be written on each overlay. Methods of measuring the scale and orientation of each photograph are described in section 3.1, and these values should be shown on the overlay, with true north indicated by a number of parallel lines used as a reference when plotting directional data.

## 2.4 Observations at the exposure

The systematic description of accurately located rock exposures is the first objective in field mapping, and involves recording data both on the map and in the field notebook. The observations normally include descriptions of the rock types present (lithology) together with both descriptions and measurements of the structures within these rocks. Field sketches and photography are useful to illustrate structures, and at each exposure you should consider whether it is necessary to collect specimens for more detailed examination in the laboratory. Stand back and take a general look before getting down to the details of measurement, description and sketching.

At each exposure, you should record in pencil *on the map* the location point, locality number, an abbreviation for the lithology, and when possible one or more symbols showing the nature and attitudes of the structures present. These observations can be inked in at base camp later that day. *In the notebook* are recorded: the locality number; the location by grid reference (if a topographic map is being used) or by reference to the aerial photograph overlay on which it occurs; a full lithological description; descriptions and measurements of structures; field sketches; a note of any photographs taken; and the number and nature of any samples collected. All this can take a good deal of time, commonly over half an hour on a good exposure. When first learning to map, or familiarising yourself with a new area, it is best to take plenty of time over each exposure and make sure nothing is missed. With more experience it is sensible to select only the best exposures for detailed treatment, and deal more briefly with the others. However, each exposure seen should be recorded on the map, if only by an unnumbered locality point and lithological abbreviation. In making a detailed map, you should see every exposure but concentrate your attention on those which show changes in lithology or structure. When reconnaissance mapping, it will generally be necessary to select exposures that lie on or near the line of traverse, but the pattern of traverses should be carefully planned to cross well exposed, key areas.

The sections that follow discuss in more detail each step in the full description of good exposures.

*Location and size of exposures.* An exposure that has not been accurately placed on the map is generally of little use, but beginners often have some difficulty with location. The best advice we can give is to consult the map or aerial photographs so frequently, whether travelling by car or on foot, that you know, at least approximately, where you are *all* the time. Then, when an exposure is found, its precise position can be determined in relation to nearby watercourses, trees, or other minor features visible on the aerial photograph. Although less satisfactory, a location may be pinpointed by taking compass bearings on distant features that can be identified on the map or photograph (see section 3.2). However, this method is particularly liable to inaccuracy when used with aerial photographs, because of the scale distortions that these contain. Another method to use when uncertain of location is to make the necessary geological observations first, then take a compass bearing on the nearest object identifiable on the map and measure the distance to it by pacing (or, when working on a small scale, with the help of the car's mile/kilometre recorder). However, the objective of this discussion is to persuade you that by far the quickest method of location is to follow the map very carefully while travelling.

Having located the exposure, the next step is to show its approximate size and outline on the map. In many cases this is easy because it is visible on the aerial photograph. Very small exposures can be shown as a point or small circle. When colouring in the field map at the end of the day, the areas of exposed rock are shown in deeper shades than are the unexposed parts of the outcrop. In this way the field map can show both places where the rocks can be seen and the evidence on which the posttioning of boundaries is based.

After marking the exposure on the map, record in the notebook its quality and general nature (cliff, gully, stream bed, artificial cutting, etc.). If the exposure is an unusually good one, or shows particularly important features, it may be worth adding notes on how to find it.

*Lithology.* In general, only very brief comments on lithology can be written on the map, most often only an indication of general character (e.g. 'bi.gn.', for 'biotite gneiss'). A much fuller account should be given in the field notebook covering, for a good exposure, such features as weathering, appearance of fresh surfaces, mineralogy, textures, and primary structures. This lithological description may be either combined with or kept distinct from the description of secondary structural features (see the following section), but in any case it is wise to remember that *both* lithology and structure must be carefully observed and recorded.

Since the approach to lithological description depends partly on whether the rocks concerned are igneous, sedimentary, or metamorphic in origin (while some ore deposits require different treatment again) we have compiled check-lists of some of the more important lithological features that should be looked for in these major rock classes (see Table 1). These lists should not be used uncritically - for each project you should compile your own check-list of features relevant to that particular study. Values should be quantified wherever possible; for instance, when describing grain size do not state only that the rock is 'coarse', or 'medium', or 'fine' in grain, but give figures for the estimated average grain size and for size variation.

When mapping a large outcrop of relatively uniform rock, there is no need to repeat full lithological descriptions at every exposure. It is normally sufficient to describe a rock type at length when it is first encountered and then describe it fully again only where it shows a notable change in character. On the other hand, beware of overlooking small but possibly significant variations in the common rock types of an area. One sometimes finds, on returning from the field and attempting to write the report, that the notebook contains numerous descriptions of all except the most abundant rock types! Make a point of checking for this kind of mistake before leaving the area.

*Structure.* Very often, the most important structural measurement made at an exposure is of the attitude of a primary structure such as bedding or flow-banding, on which the strike and dip should be recorded wherever possible (for methods of doing so see Section 3.3 and Figs 11 and 12). In addition, in many cases there will be secondary structural features to describe and measure, such as cleavage, jointing, folding and tectonic lineations. When examining strongly deformed rocks, students sometimes find it difficult to distinguish between bedding and cleavage, or between bedding and a strong jointing direction. Where this is so, a careful examination will usually reveal that the bedding is picked out by variations in colour or grain size, or by primary sedimentary structures such as cross lamination (see Fig. 5a, for example). Structures should not be described as 'bedding', however, unless there is evidence that they are indeed of primary sedimentary origin. For example, the banding common in gneisses should never be recorded as 'bedding', but as 'lithological banding'.

In Table 1 and Figs 2-8 and 13, we have listed and illustrated some of the structural features that deserve proper description and measurement wherever they are found. In this compilation we have included only the more straightforward features; for detailed descriptions of structural features consult a good text (see the Bibliography). We wish to draw special attention, however, to the importance of searching for **way-up** (or **geopetal**) structures in areas of highly folded, bedded rocks. Figure 8 illustrates a number of structures which can be used in this way, although unfortunately some of these are rare in rocks older than the Cambrian. In regions of strong folding it is wise to check the way-up of beds at regular intervals, even if the dip has not changed significantly.

**Table 1** Check-lists of lithology and primary structures. These lists are intended only as reminders of important features which should be looked for, and properly described if found: no attempt is made here to discuss their nature and origin.

| Feature | Comments and useful terms |
|---|---|
| *In sedimentary rocks* | |
| colour | When fresh and when weathered. |
| weathering character | Relative resistance of various beds; special weathering characteristics (e.g. honeycombed, carious, box-worked). |
| strength | Loose, friable, plastic, soft, weakly cemented, compact, hard, brittle, etc. |
| density | Give an estimated value (g cm$^{-3}$). |
| mineralogy | List identifiable major and minor constituents (with approximate percentages); distinguish mineral grains from lithic grains, clastic grains from cement, phenoclasts from matrix; use the dilute acid test for carbonates. |
| texture | In clastic sediments observe: grain size, its variations, sorting (a standard scale may be made from sieved samples glued to a card and labelled); grain shape, roundness, sphericity; fabric characteristics (e.g. grain-supported, mud-supported, pebble orientation, depositional lineation, imbricate texture, concretions, porosity). In chemical/biochemical sediments, observe such features as grain shape, grain size, fabric, oolites, pisolites, concretions, porosity. |
| fossils | Distinguish trace fossils and body fossils; their distribution within the sedimentary units (e.g. uniform, dispersed, in layers, in lenses); general identity and relative abundance of the various types; mode and state of preservation; entire or broken; in-life position or transported; ecological associations; record any evidence of bioturbation (e.g. mottled or homogenised beds) and, for burrows, describe infill as well as matrix. |

| primary structures | Describe bedding characteristics such as: thin- or thick-bedded (give dimensions); massive or laminated; continuous or lensoid; plane bedding; cross bedding; graded bedding. |
| --- | --- |
| | Other primary structures may include: wash-outs; ripple marks; sun-cracks; rain prints; flute casts; groove casts; prod marks; load casts; flame structures; clastic dykes; slump structures; bioturbation. |
| local successions | Well exposed sections in gullies, cliffs, etc. should be carefully measured up, recording bed by bed the details of lithology and structure. |

## In volcanic igneous rocks

The check-list for sedimentary rocks is also applicable in many respects to volcanic sediments, although some of the special terms below are also needed. Lava flows require different treatment.

colour, weathering character, strength, density

| mineralogy and texture | Describe as for intrusive igneous rocks in the case of lavas, and as for sedimentary rocks in the case of volcanic sediments. |
| --- | --- |
| in air-fall pyroclastic rocks | Bombs, blocks, cinders, lapilli, crystals, ash, pumice, bomb-sags, agglutination. |
| in pyroclastic flow deposits | Pumice, lithic fragments, lapilli, ash, crystals, welding (and its variation), eutaxitic foliation, pumice fiamme, cooling joints, lithophysae (i.e. siliceous nodules), flow units, cooling units. |
| in reworked pyro-clastic material | Record the relative amounts of volcanic and non-volcanic material. |
| in basic to inter-mediate (low viscosity) lavas erupted sub-aerially | Scoriaceous tops and bases, massive centres, vesicles, amygdales, weathered tops, 'pahoehoe' and 'aa' surfaces, columnar cooling joints, platy flow textures and jointing. |
| in low viscosity lavas erupted sub-aqueously | Pillow structure; record pillow size, internal zonation, vesicularity, cooling joints, related hyaloclastites and other sediments. |

| in high viscosity lavas (usually acidic to intermediate) | Autobrecciation, blocky surfaces, ramp structure, flow-banding, flow-lineation, cooling joints. perlitic cracks, hyaloclastites, spherulites, vesicles. |

## In intrusive igneous rocks

| colour, weathering character, strength density, mineralogy | Include colour index, and distinguish between primary minerals and alteration products. |
| texture | Distinguish between coarse-, medium- and fine-grain; equigranular, inequigranular, seriate and porphyritic textures; xenomorphic-granular, hypidiomorphic-granular, panidiomorphic-granular textures; random or preferred orientation (planar or linear); banding. |
| contacts | External (i.e. with country rocks) and internal; concordant or discordant; sharp, gradational, veining, straight or sinuous, chilled or unchilled. |
| xenoliths | Composition, size, shape, sharp or gradational contacts, *schlieren*, abundance. |
| aplites, pegmatites, miarolitic cavities | Dimensions, abundance, mineralogy, texture. |
| joints | columnar, cuboidal, sheeted, tensional, shear, mineralised. |

## In metamorphic rocks

colour, weathering character, strength, density, mineralogy

| textures and structures (i.e. fabric) | Grain size and its variation; porphyroblasts and matrix. |
| | Hornfelsic, granoblastic, slaty, phyllitic, schistose, cataclastic, flaser, augen, mylonitic and gneissose fabrics. |
| | Relict features (e.g. bedding, pebbles, amygdales, phenocrysts). |
| | Planar structures (e.g. planar preferred orientation of minerals, banding, cleavage, jointing, shearing. |
| | Linear structures: linear preferred orientation of minerals, plane intersections, crenulations, slickensides. |

See Figures 2-8 for additional structural features.

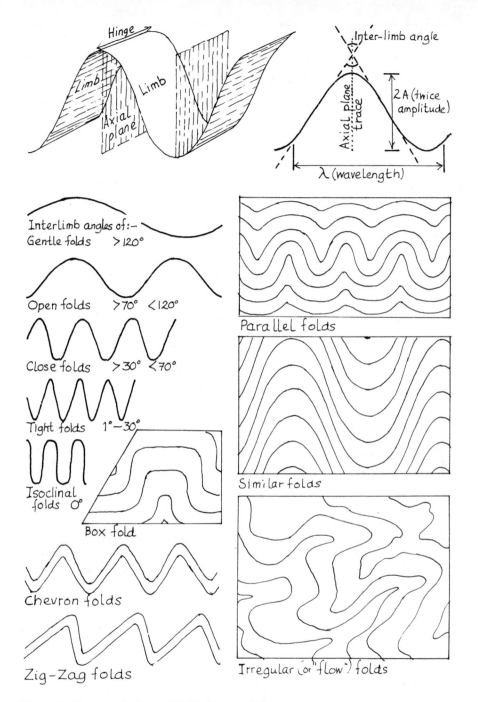

**Figure 2** The terminology of fold shape and size.

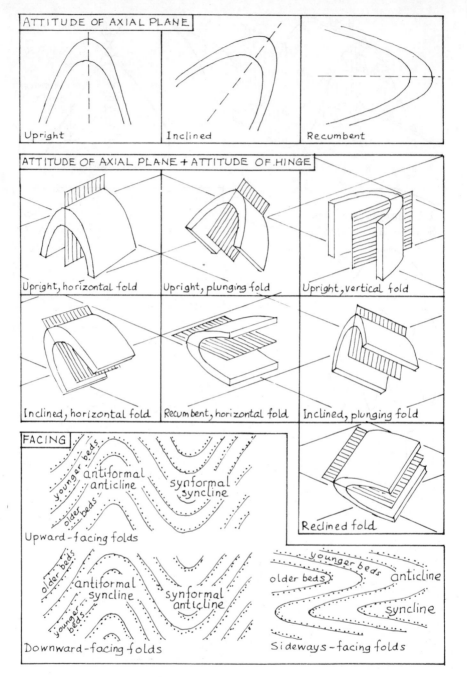

**Figure 3** The terminology of fold attitude.

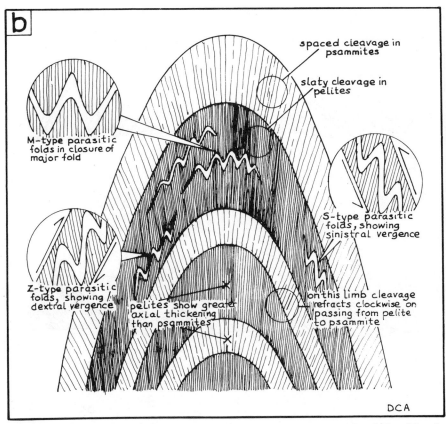

**Figure 4** (a) Types of cleavage. (b) Some common structural relationships in folded mixed sediments. Note that for large folds the positions of hinges and the directions in which they close may be predicted from small exposures on the limbs which show cleavage/bedding relations or parasitic Z and S folds (see also Fig. 5).

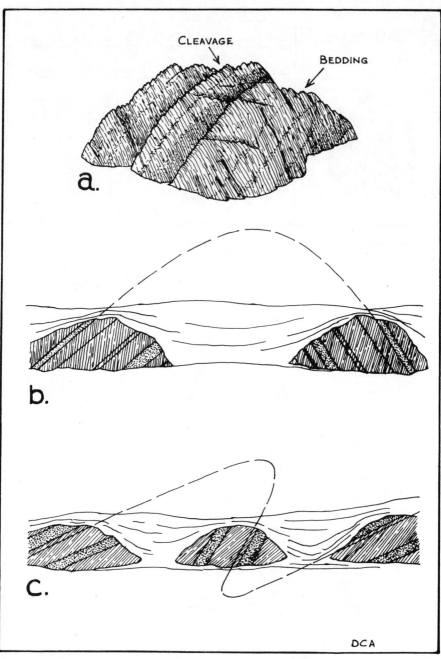

CLEAVAGE

BEDDING

a.

b.

c.

DCA

**Figure 5** Using the relationship between bedding and axial plane cleavage. (a) Note that cleavage is often as prominent as (or more prominent than) bedding - to distinguish bedding look for planar variations in features such as grain size and colour. (b) An example of how the bedding/cleavage relationship can be used to indicate the position of a fold hinge and to distinguish whether it is antiformal or synformal (compare with Fig. 4). (c) Note here that the bedding/cleavage relationship in the middle limb of this antiform/synform pair shows this limb to be overturned. If the structure as a whole is right way up, the beds in this middle exposure must young downwards and to the right. On the other hand, the whole structure might be upside down, in which case the beds in this exposure will young upwards and to the left. Thus the bedding/cleavage relationship should not be used independently as a 'way-up' structure.

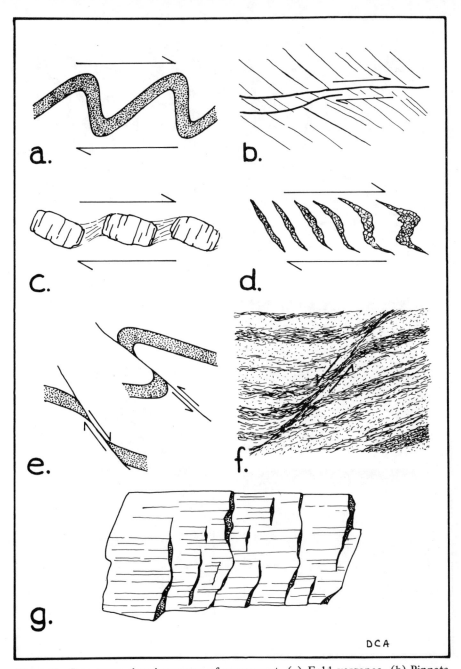

**Figure 6** Structures showing sense of movement. (a) Fold vergence. (b) Pinnate tension fractures adjacent to a fault. (c) Rotated boudins seen in cross section. (d) En echelon quartz-filled tension gashes (straight on the left, sigmoidal on the right). (e) So-called 'drag' folds adjacent to faults. Such folds may precede or accompany fracturing and should be interpreted with care since they occasionally show the opposite sense to the fault. (f) Drag associated with a shear zone in gneisses. (g) A stepped slickensided surface like this is often the result of movement of the nearer fault block to the right, i.e. the steps face in the direction of movement of the opposing block. However, experiments have shown that in some cases such steps can arise during movement in the reverse direction, so confirmatory evidence should be sought.

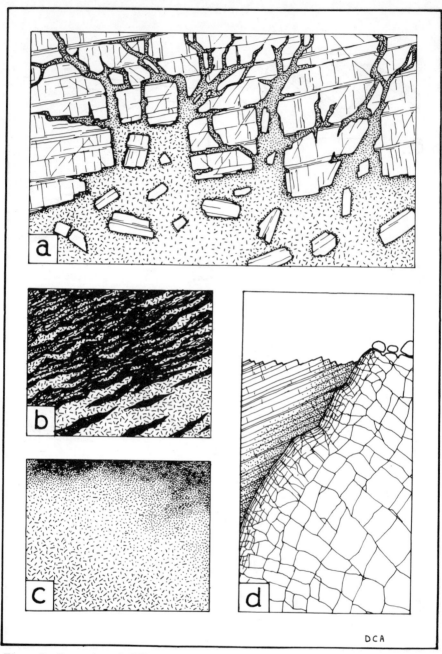

**Figure 7** Examples of different styles of intrusive igneous contact. (a) Irregular, sharp and cross cutting; the igneous rock shows chilling against the country rocks and penetrates them as fracture-controlled veins; the numerous xenoliths of country rock within the intrusion appear to have been enclosed by stoping. (b) Unchilled, conformable contact showing a discontinuous gradation into the country rocks through a zone of migmatites. (c) An irregular contact which shows more or less continuous gradation from igneous rock into country rock. (d) A sharp, smoothly regular and cross cutting contact which shows chilling and flow-banding in the marginal zone, and cooling joints which become increasingly widely spaced inwards. The country rocks display a narrow zone of contact metamorphism.

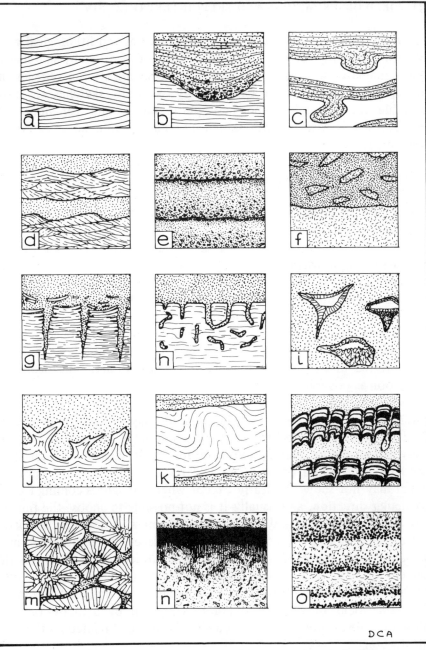

**Figure 8** Way-up (or 'geopetal') structures: (a) truncated cross bedding or cross lamination; (b) channelling; (c) load casts; (d) ripples; (e) graded bedding; (f) the upper bed contains clasts of a lower bed; (g) sand-filled mud cracks and saucer-shaped mud flakes; (h) a surface bored by burrowing organisms; (i) part-filled vugs in a sediment (or part-filled amygdales in a lava flow); (j) flame structure; (k) slump structure, here truncated; (l) in situ organic growth, in this case of an algal colony; (m) in pillow lavas the lower surfaces of higher pillows mould over the pillows below; (n) the weathered upper surface of a sub-aerial lava flow; (o) density grading in layered basic igneous rocks.

*Field sketches and photographs.*   The structural and lithological com-plexities presented by some exposures are best recorded by sketches or photographs as well as by written notes. Drawing also encourages the geologist to look at the rocks carefully. Accurate sketches are more valuable than any but the best photographs, and for important exposures it is wise to both sketch *and* photograph. Some points to remember:

(a)   Make sketches large and clear (often use a whole page of the field notebook).
(b)   Label all parts of the drawing and show the exact places from which any samples were collected.
(c)   Always include a scale on a sketch and an object of known size within a photograph.
(d)   Draw structures such as folds in profile (i.e. cross section) and show on the sketch the orientation of the drawing (e.g. 'cross section of folds as seen looking downplunge to NE').
(e)   The location, direction (e.g. towards NW) and subjects of all photographs should be written into the field notebook and the printed photographs should be indexed with area and locality as soon as possible.

Either colour or monochrome photography may be used for geological field recording. Monochrome prints sometimes show details more clearly than colour photographs and are much cheaper and easier to produce. One major problem in field photography is that important detail may not show up clearly except when the light is coming from a particular direction, and usually it is not possible to choose the time of day for one's photography. This is one reason why drawings are often more useful than photographs. In shaded places, the photograph may sometimes be improved by using a flash attachment, either on the camera or held to one side in order to highlight relief on the rock surface. Stereoscopic effects may be obtained by taking photographs of the same field of view from points separated by a metre or two. Moseley (1981) offers a good discussion of this, and related methods of illustration.

*Collecting samples.*   In many mapping projects, the main purpose of taking rock samples is to obtain a collection of hand specimens repre-sentative of the rock types and possibly also of fossil material to be found in the area. Samples may also be taken to allow more precise identification of rock types. Hand specimens are useful both in themselves, as a record, and as a source of material for various laboratory techniques - thin sections, peels, tests for density, porosity, permeability and mechanical strength, and material for mineral separation and chemical analysis. For the simpler tests, including thin sectioning, specimens of about 0.5-0.75 kg are sufficient, although larger specimens may be necessary when the grain size is very coarse or the rock is being collected for the examination of structural features. A hand specimen size of about 10 cm x 6 cm x 3 cm is of roughly this weight and a convenient shape for storage. However, if chemical analysis, mineral separation or mechanical tests are intended, the minimum specimen size will have to be 1.5-2.0 kg. For geochronological work even more material will be needed, the amount depending on the method to be used, and advice should be sought from the geochronologist who will be carrying out the determinations.

Ideally, each sample in the collection will be either of completely fresh material or fresh except for one surface that shows its weathering characteristics. However, it will sometimes be necessary to collect less than ideal material because nothing else is obtainable.

The field geologist has to take care that his collection is not too strongly biased towards the less common rock types. It is normal for this bias to occur, and with good reason, but it should be partly balanced by a conscious effort to collect a number of good specimens of the more common types. Another bias in most collections is towards the rock types which are most resistant to weathering and erosion, and so likely to be exposed most often.

Because of the various biases of normal 'spot sampling', a more carefully organised sampling scheme will be needed if the purpose of the project is to collect material truly representational of areal distribution. For instance, if the purpose is to measure chemical variation in a single igneous intrusion, samples might be collected at regular intervals along evenly spaced traverse lines, or at the intersections of a regular grid. Unfortunately, exposures are rarely good enough to allow such a scheme to be completely fulfilled. As another example, one common method of determining the average metal content of a tabular ore body is to chisel out a 'channel sample' about 3 cm deep and 9 cm wide, across the entire width. All the material removed is combined into a single sample, which is crushed, thoroughly mixed and split into small amounts for analysis.

For studies of structural or magnetic properties, it is often necessary to collect 'oriented specimens'; that is, rock samples which are marked to show how they were spatially positioned when collected. One way of doing this is to hammer off a suitably sized specimen and then fit it back into its original position on the exposure while marking it up (Fig. 9c). For instance, it may be marked with a horizontal line (passing round three or four sides) and with an arrow pointing north. An alternative method is to measure and mark the strike and dip on any plane surface on or running through the specimen, and also indicate the top of the specimen. Such marking up is best done with a felt-tip pen.

Specimens must be labelled at the time of collection with a reference number corresponding with that marked on the map and in the field notebook. Again, one quick method is to use a felt-tip pen; another is to mark the number on a piece of thick self-adhesive tape (e.g. Elastoplast tape) attached to the specimen. The specimen number can be marked on more permanently in camp or in the laboratory by the application of a patch of quick-drying enamel on which waterproof ink can be used. To protect specimens during transport back from camp to laboratory, it is useful to have a supply of newspaper and wooden boxes into which the wrapped specimens can be tightly packed. Easily broken rocks, minerals and fossils are best wrapped in soft paper and placed in small containers (tins, cardboard boxes) before being packed.

1 *While moving from one exposure to the next consult your map frequently in order to know your approximate position at all times.*
2 *The map should be marked* in the field *with symbols showing location and size of exposure, general lithology and main structures, as well as with lines showing topographic features and possible boundaries.*
3 *In the notebook should be entered a* field *record of location, detailed lithology and structure, sketches and sketch maps, and notes of any photographs or samples taken.*
4 *Use a field check-list as an aid to thorough observation.*
5 *Use a standardised system of symbols and abbreviations.*
6 *Quantify values whenever possible.*
7 *Good field sketches are of great value, and often more useful than photographs.*
8 *Samples can be used to aid identification, to provide a record of the rock types in the area, or as material for subsequent laboratory studies. Make sure they are large enough and fresh enough for their purpose, and beware of potential sampling bias.*
9 *Ink in your map and check your notes at the end of every day.*

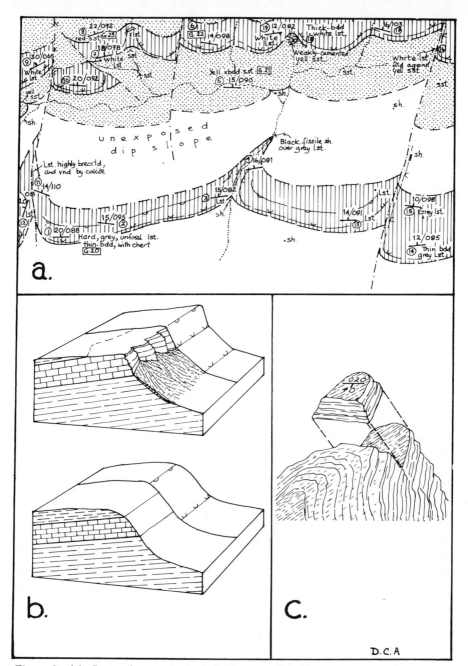

**Figure 9** (a) Part of a geological field map. (b) Two common relationships between breaks of slope and the gently dipping sheets of resistant rocks which caused them - note the positions of the contacts relative to the breaks. (c) One method of collecting an oriented sample.

## 2.5 Geological boundaries

*The mapping unit.* In the previous section we described the kind of observations which should be made at each exposure of rock. In this section we discuss the mapping in of the outcrops, of which the exposures form only a part. Whereas the work at individual exposures is mainly a matter of observation, this next stage requires a good deal of interpretation.

It is necessary to decide at an early stage how the various rocks of the area may best be grouped into mapping units, each of which can be shown as a separate outcrop on the map. Later, the field evidence collected both from the exposures and from the unexposed ground between them is used to infer the probable positions of the geological boundaries between the units. All interpretation should be done while in the field, with the evidence still available to be checked as necessary. Normally, the mapping in of boundaries is done as opportunities occur in the course of exposure mapping. Interpretation should not be all left until the end, but carried out while the evidence is still fresh in your mind.

The first problem, then, is to select suitable mapping units. A mapping unit may be defined as a body or rock which is distinctive enough to be mapped separately from adjacent rocks, and large enough to be adequately shown at the base map scale. The distinctive character of a mapping unit usually depends on its lithology, which may be uniform or may vary within defined limits. Sometimes it is based on other characteristics, such as fossil content for example. As to size, it is useful to remember that, at a scale of 1 : 10000, a thin pencil line (0.2 mm wide) represents a strip of ground about 2 m wide. Any outcrop narrower than this must either be exaggerated or not shown on the map. Clearly, the smaller the mapping scale the more likely is the need for some exaggeration. So whereas the outcrop of every mapping unit should ideally be shown at true scale, in practice a certain amount of scale exaggeration is allowed, especially in the case of small, but significant, rock bodies such as an igneous dyke or a thin sedimentary layer within a succession of lava flows.

In an area which has been mapped previously at a similar scale, the geologist must decide whether he will use the same mapping units as his predecessor or define new ones. In this case he will have, from the beginning, some conception of the kind of lithological features on which the division into units can be based. On the other hand, if the area has not been mapped before, the selection of suitable units must be made as the work proceeds, and some early ideas may have to be revised in the light of later experience. Many decisions on the selection of units are, however, easy to make. Thus, there can be no doubt about the need to map as a separate unit a thick band of limestone within a succession of shales, or a stock of granite intruded into schists. More difficult decisions arise when the rocks either vary in a complex way or vary only gradually over considerable distances. For example, it may be difficult to select mappable boundaries in a sequence of rapidly alternating sandstones and shales, or in a region of high-grade quartzo-feldspathic gneisses. In some cases the identity of the most significant mapping units may become clear only after the geologist has mapped a large part of the area in detail.

Commonly, a beginner will attempt to select mapping units which are too small to be traced successfully across the whole area. It is rarely possible to show on the map more than a generalised picture of the observed lithological variation, because of the limitations imposed by the amount of exposed rock and the scale of the mapping. Good field maps will generally show a great deal of local variation *within* each mapping unit. An important part of the geologist's work is to use good judgement in making the most practical and significant generalisations.

Occasionally, however, one finds a unit which, although small, is so distinctive in character and extends over such a wide area that it is of the greatest value in correlating successions. Such marker horizons should be looked for and carefully mapped in wherever found. Beds of coal, thin limestones, and distinctive volcanic tuffs have often been used in this way.

Eventually, each mapping unit should be given a proper name, and when the units form part of a clearly defined stratigraphical succession, the rules of formal lithostratigraphical nomenclature should be followed. Most importantly:

(a)   A 'formal' name should be given initial capital letters and refer to a specific place or region (e.g. Atbara Sandstone, or Bayuda Volcanic Group).

(b)   The Formation is the division most often used for the individual mapping unit. Units selected as formations should be (a) large enough to be mappable with little or no exaggeration of scale, (b) show a large degree of lithological homogeneity, and (c) be clearly separable from adjacent formations. The term itself may or may not be incorporated into the formal name (e.g. Jebel Atshan Sandstone *or* Jebel Atshan Formation would both be correct as formational names).

(c) Lithologically variable formations may be divided into Members, and more rarely it may be useful to divide a member into Beds. However, the term 'bed' (without initial capital) is more often used as an informal term.

(d) Two or more adjacent and related formations may be referred to as a Group, while the term Supergroup is used for two or more adjacent and related groups.

There are sometimes problems in applying the formal lithostratigraphical terms to igneous and metamorphic rocks, but the principles remain the same. Volcanic successions can be treated in a similar way to sediments, but often show more rapid lateral variation. Consequently it is often best to define formations by grouping together a number of lava flows or volcaniclastic units which are related by composition or mode of eruption. For instance, one might apply a formational term to a sequence of petrographically similar basalt lavas, or to a succession composed of several ignimbrite flows. Individual flows can then be separately referred to as 'Members', when required. Large intrusions of igneous rock are named in a similar way to sedimentary formations, but associated intrusions are generally grouped under informal terms such as 'complex' or 'suite', rather than referred to formally as 'Groups'. In metamorphic terrains, formational names are best based on the broad lithological character which is normally an inheritance from the pre-metamorphic stage (e.g. Erkowit Pelite, Haiya Marble, Sabaloka Biotite Gneiss). Metamorphic zones and other features formed during metamorphism may then be shown separately, superimposed on the formations. However, it is not always possible to follow this scheme. For example, a thick zone of mylonite along a strike-slip fault must clearly be mapped as a separate formation even though it has been generated during deformation and metamorphism, and owes little of its character to pre-metamorphic stages.

For further details see the guide to stratigraphical procedure published by the Geological Society of London (1972).

*Mapping contacts.* The general positions of geological boundaries will become increasingly obvious as the mapping in of individual exposures proceeds, but it is often a good idea to make special traverses along the line of important contacts, preferably starting at a point where the contact is exposed, or at least well defined by nearby exposures. From here it should be followed along its trace marking in: (a) sections where it can be precisely located, either because it is exposed or because it can be fixed to within a few metres by adjacent exposures; (b) sections where its position can be inferred to within 10-20 m; and (c) sections where its position is conjectural. These different degrees of certainty in placing the contact are expressed on the map by varying the contact line between continuous and pecked, the peck spacing denoting the degree of uncertainty (see mapping symbols in Appendix B). A special symbol may be used for a gradational contact, which is mapped either by drawing a line along the centre of the zone of gradation, or by mapping in points at which a particular mineralogical or textural feature is first seen when traversing the contact zone in a given direction.

The boundary between any two mapping units may mark a line of normal contact (that is, where the rocks of one unit lie conformably adjacent to the other in stratigraphical succession) or it may define an unconformity, a fault, or an intrusive igneous contact. All these different types of contact have their own characteristics, which you should seek out as you map in boundaries, but they also have a number of features in common. We will consider the common features first.

The ways in which a line of contact may be expressed at the surface can be described as follows:

(a) *Changes in solid rock lithology.* If there are sufficient exposures near to it, the boundary may be sufficiently defined by taking the simplest, geologically reasonable line between exposures of the two contrasted lithologies which define the mapping units. When drawing such a line, account should be taken of the effects of topography on the outcrop shape of planar structures (e.g. the way in which beds form V-shaped outcrops where they cross valleys).

(b) *Topographical features.* When continuing boundaries through unexposed ground, topographical features are extremely useful guides to outcrop trends, and should be shown on the map wherever possible (see Appendix B for appropriate symbols). Such features can often be seen most easily and plotted most accurately while examining aerial photographs stereoscopically. When observing features on the ground it helps to look for them when the light is oblique, in the early morning and late evening. The commonest kind of feature is the break of slope, which generally marks the differing resistance of adjacent lithologies to weathering and erosion. Note that, where a scarp or a line of steeper slopes marks the outcrop of a more resistant rock type, the base of the feature-forming bed will lie some distance above the lower, concave break of slope (Fig. 9b). Watercourses and spring lines may also pick out contacts, and lines of sink-holes often mark the tops of limestones interbedded with shales or sandstones. Igneous dykes and quartz veins are often expressed as sharp, linear ridges, although in some places dykes form gullies instead.

(c) *Changes in superficial materials.* Fragments, blocks or boulders lying on or near the surface, and known as float, may have been derived from immediately underlying solid rocks, but care should be taken to consider how far they might have been transported from their parent outcrop. For example, on a scree-covered slope it is obvious that all the fragments have moved down slope from their origins by varying amounts, but it may still be possible to trace a contact across the slope by remembering that the outcrop of any unit must lie up slope of the points where fragments of the unit below can be found. It should be also understood that the more resistant rock types will have disproportionally large representation in the float. Where boundaries are obscured by a soil cover, the colour and texture of the soil material may indicate the general nature of the rock beneath, provided that the soil has not been transported into its present position. Thus sandy soils often develop over sandstones, granites or quartzo-feldspathic gneisses, whereas dark, heavy soils may cover clays, shales or micaceous schists. However, some of the common soils of Africa, including red lateritic loams and black cracking clays (or 'cotton soils'), reflect little of the nature of the rocks beneath, the first being essentially the product of a particular climatic region, while the black clays have generally been transported. Differences in soil properties in turn largely control the character of vegetation, so that changes in plant cover may also give clues to the positions of unexposed boundaries.

*Unconformable contacts.* If this type of contact is exposed it may show an angular discordance in attitude between the beds above and below it. Indeed, even if the contact plane itself is not to be seen, such discordance is often detectable from the dips and strikes on exposures near to it. On the other hand, some unconformities are disconformities and lack angular discordance.

An unconformable contact surface may be irregular in shape owing to the presence of palaeo-relief, with the upper series burying the hills and valleys of an ancient landscape. The scale and character of the buried relief should be carefully recorded in such a case. In the absence of angular unconformity or paleo-relief, the presence of an unconformity may be detected by the occurrence of one or more of the following features:

(a)    A rapid, upward change in lithology, often directly from contin- ental sediments into marine, or the reverse. There may be a well developed conglomerate at the base of the upper series.

(b)    The truncation of faults, dykes, or larger igneous intrusions by the beds of the upper series.

(c)    Fossil evidence of a long time gap between the sediments above and below the contact (shorter intervals may be marked by the presence of bored surfaces or epifaunal encrustations.

(d)    The presence of a weathered surface, phosphatic layer, or palaeo- sol at the top of the lower series.

*Fault contacts.* Fault lines are often picked out in the topography as long, almost straight linear features which may be marked by breaks of slope, valleys, watercourses, or spring lines, following alignments often, but not always, oblique to the local structural trend. Such features are often most clearly seen on aerial photographs, but are there easily confused with the similar lineations produced by master joints.

Another common characteristic of faulting is the offsetting of beds, or of structures, when these are followed along strike into the fault zone. Again, this is often well seen on aerial photographs, but faults which strike parallel to bedding cannot show offsets of this kind.

Other indications of faulting may include:

(a)    The repetition or omission of beds by the fault displacement. This is particularly common where the fault strikes more or less in the same direction as the bedding, and is especially important in the case of slides.

(b)    The presence of structures such as slickensides, drag folds, or quartz-filled tension gashes.

(c)    The occurrence of cataclastic rock types such as fault breccias, phyllonites, sheared gneisses, or mylonites.

(d)    Zones of strong silification, quartz veining, strong alteration, or mineralisation.

Some measure of the *apparent displacement* on a fault may be obtained if it is possible to match up either stratigraphy or structures across the fault line, but any measure of the *true displacement* is usually very difficult to obtain.

The *direction of movement* of a fault can often be estimated from slickensides, or from the attitudes of tension gashes and folds related to the faulting (Fig. 6). Thus, movement direction may be given by the orientation of striations or grooves ('slickenlines') on a slickensided surface, and is measured as for other linear structures (see Section 4.4). The *sense of movement* (left- or right-handed, up or down) may also be revealed by related folds, tension fractures or, in some cases, by the small steps which are common on slickensided surfaces (Fig. 6). However, experimental evidence suggests that this last criterion is not entirely reliable and, wherever possible, more than one line of evidence should be sought. Remember also that many faults, especially large ones, have moved more than once during their history, and the direction or sense may have been different in each phase of movement.

*Igneous intrusive contacts.*   The major problem in mapping intrusive contacts is that they are less predictable than normal contacts or faults. An intrusive contact which in one place is a regular, planar surface may in another become highly irregular in form, and consequently produce a complex outcrop trace. To define such contacts accurately they must be followed in detail, metre by metre, assuming that there are enough exposures to allow this to be done. It is true that dykes and sills are relatively predictable in behaviour, since they are largely controlled in form by pre-existing joints and bedding planes, but even these intrusions are liable to step abruptly from one structural plane to another. Since dykes have a strong topographic expression, their courses are often best followed on aerial photographs.

Sills present the particular problem that they are not always easy to distinguish from lava flows. The most important criteria used in making this distinction concern the relationships seen at the upper contact of the igneous sheet:

(a) The top surface of a lava flow (but not of a sill) may show palaeo-weathering or erosion, and in rare cases the sediments may contain fragments of the underlying lava.
(b) A sill (but not a lava flow) may cause weak contact alteration of the immediately overlying sediments, and may vein its roof.
(c) The tops of lava flows may be highly vesicular, even scoriaceous, or blocky, sometimes with convoluted flow-banding. The tops of sills are normally more regular and contain few or no vesicles.
(d) Inclusions of the overlying beds are occasionally found in sills, but never in lava flows.

When mapping intrusive contacts, it is important to measure contact attitude (dip and strike) wherever possible, as well as noting features such as chilling, veining, inclusion of country rock, and contact metamorphism. Proper information on contact attitude is critical to interpretation of the *shape* of the intrusion and consequently to its *mode* of intrusion. In this respect, a record of the attitude of the contact, as seen on a large scale in profile on hill slopes, is often more useful than measurements on small exposures of the contact plane itself. On the other hand, in determining the relative ages of intrusions, detailed observations of veining, chilling, and the enclosure of one rock type by another are essential. Occasionally, however, veining relationships can be deceptive, as when a hot basic magma has invaded an earlier intrusion of acidic rock and caused partial melting of it near to the contact. When the basic rock has cooled it may have been veined by the remnants of molten acidic material ('back-veining'), giving the incorrect impression that the acidic intrusion is the younger.

1  *Tracing important contacts along strike is a complementary method to making traverses across strike, but should be used sparingly.*
2  *Insert contact lines in the field, as the mapping progresses.*
3  *Indicate degree of certainty along a mapped contact by means of continuous to broken lines.*
4  *Differentiate between normal (stratigraphically conformable) contacts and contacts which mark unconformities, faults, or the margins of intrusions. Indicate in your notes the evidence upon which such distinctions have been made.*
5  *Igneous intrusive contacts often require special attention.*


## 2.6  Mapping in forested areas

Forests and heavily wooded areas offer special problems of accessibility, location and the finding of exposures. Savannah vegetation can also be difficult during the long-grass season. Mapping rates can be much lower than those mentioned in Chapter 1, and the whole process is less efficient since fewer exposures are seen. Under these conditions, aerial photographs are invaluable in defining the drainage system, the location of bare rock, and the presence of any breaks of slope likely to have exposures on them. Even in heavy forest the tree canopy can closely reflect the form of the ground surface below, so that all but the smallest topographical features can be picked out. If, in addition, a contour map is available it can be used in conjunction with an altimeter as an aid to location.

Watercourses assume particular importance in forested country since they may provide access routes, have exposures along them, and are a great help in accurate location. Consequently, the drainage pattern may form the basis of a traverse network. Even so, some lines cut through the intervening bush may be necessary to fill in gaps and for vehicle access. Since cut lines are expensive and time consuming to construct, their use as the main element in a traverse network can only be justified over small areas and where there is known economic potential. In such cases the lines are cut parallel to each other on a fixed bearing laid out at right angles to the lithological strike of the rocks. The line spacing should be less than the minimum expectable strike length of the type of deposit being sought.

## 2.7   Superficial deposits

Although in this book we are concerned mainly with the mapping of solid geology, the character, distribution and thickness of superficial deposits encountered on a mapping exercise cannot, and should not, be ignored. This is particularly so where these materials are thick, extensive, or of economic importance. Such deposits can be shown on the map either as separate formations with their own ornament or colour, or within lightly defined boundary lines laid over the solid formation ornaments. A scattering of symbols can then be used to indicate either their origin (i.e. dune sands, alluvium, lacustrine sediments) or their general composition. Note that if the primary objective is to map solid geology, the presence of obscuring superficial deposits is not an excuse for leaving gaps in the boundary lines of the underlying solid formations; the boundaries should be continued beneath the superficial cover as broken lines indicating their conjectured positions (Fig. B1). You should in any case record, on the field map, areas of 'no exposure' of solid geology, provided that the absence of exposed rock has been checked on the ground. By this means one can distinguish between areas examined unsuccessfully for rocks, and areas which may appear on aerial photographs to lack exposures but which have not been ground checked. In the field notebook, as well as on the map, an attempt should be made to distinguish the main lithological varieties - sands, silts, clays, gravels, thick organic-rich soils, laterites and so on. As remarked elsewhere, the character of the 'float' can also provide clues on the nature of the underlying solid rocks if the superficial deposits have formed in place, or on the character of solid rocks exposed upstream in the case of a transported deposit.

## 2.8   Geological interpretation in the field

A field mapping project should progress at three different levels more or less simultaneously. The first level involves the careful and systematic description of the rocks observed at individual exposures. At the second level, information from the exposures is used to divide the rocks of the area into a number of mappable units and insert boundaries between them, using additional evidence from topographical features and super-ficial deposits. This stage inevitably involves a certain amount of geo-logical interpretation. The third level is entirely interpretative and uses the conclusions reached at the other two levels to propose general hypotheses about the geology of the area. Some of these hypotheses will be concerned with the recognition of distribution patterns; for example, the structure of the area as determined by the distribution of folds, faults, cleavages and so on; the pattern of mineralisation; the distribution of metamorphic zoning. Other general hypotheses will attempt to explain the geological history that is responsible for these distribution patterns - sequences of sedimentary and faunal facies; structural and metamorphic histories; the sequence of igneous activity. The question arises as to how much of this interpretative geology can and should be done in the field, as against its deferment for later consideration in the office. Clearly it is not possible to reach final conclusions in the field if some of the evidence will only become avail-able after laboratory examination of collected material. For instance, important decisions may rest on the precise identification of particular fossils, or on the chemical characteristics of a suite of igneous rocks. Nevertheless, there is one very strong argument for reaching at least provisional conclusions while still in the field. This is that, by doing so, one still has the opportunity to test the hypotheses against additional field evidence and to collect further specimens of critical material. The process of creating a hypothesis almost inevitably suggests ways in which the evidence could be cross checked, and if this stage is not reached until later, in the office, it may be very difficult or impossible to carry out these checks. This is particularly true when the hypothesis is concerned with, or depends upon, structural evidence, and every effort should be made to reach an advanced stage in the structural interpretation while still in the field. In particular, before leaving the field you should construct draft versions of the cross sections which are to accompany the final map, and perhaps make preliminary versions of stereographic plots, and other graphical and statistical displays of the data.

# 3 Field measurements

## 3.1 Determining the scale and orientation of aerial photographs

Aerial photographs are taken in continuous sequences or 'runs' while the aeroplane is being flown on a set compass bearing (often N-S or E-W). Therefore, all the photographs in a run have a similar orientation. Nevertheless, there is significant variation from photograph to photograph, making it necessary, in the absence of a good topographical map, to determine the precise orientation of every print which is used as a base for mapping. Similarly, although the pilot attempts to keep the aeroplane at a constant altitude there are always variations in scale due to changes in height above ground. However, scale variations within a run are normally less significant than the variations in orientation, and it should be sufficient to determine scale on only a few photographs within the mapping area.

When selecting an area in which to determine scale and orientation of an aerial photograph, remember that distortions are smallest in the central part of the photograph and in areas of low relief. Having found a suitable area on the photograph proceed as follows:

(1) *Choose two points* which are visible from each other and can be easily identified on the ground, and which are about 5 cm apart on a standard size aerial photograph. This separation represents about 1.25 km distance on a photograph at 1 : 25 000 scale, and 2 km at a scale of 1 : 40 000.

(2) Travel to one of the points and *take an accurate compass bearing* on the other (Fig. 10a-c). Whenever using the compass, keep any heavy steel or iron objects (such as a hammer) *well away* from it.

(3) Travel to the second point and *take a back-bearing* on the first to check accuracy. From the known direction of the line through the two points set off the direction of north (either magnetic or true). Draw parallel north-south lines in pencil across the whole photograph overlay, for use when plotting bearing and structural symbols (Fig. 10d).

(4) *Measure the distance* between the two points on the ground and divide by the distance separating them on the photograph measured very carefully) to determine scale. When calibrating small-scale photographs, the ground distance can be measured with

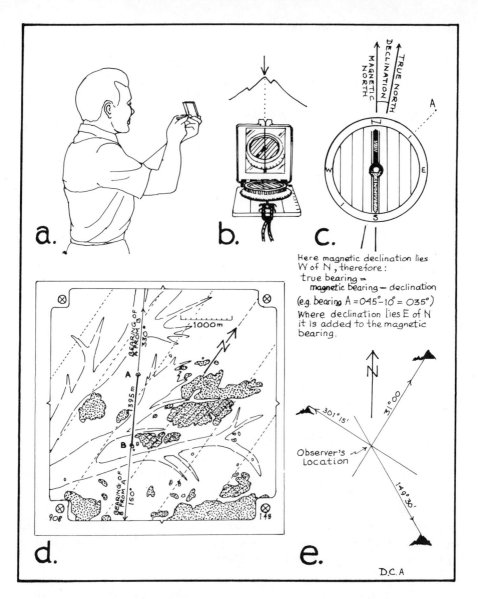

**Figure 10** Using compass bearings: (a) taking a bearing on a distant object; (b) how the object is aligned with a Silva compass; (c) the correction of magnetic bearings to true bearings; (d) marking orientation and scale on an aerial photograph overlay; (e) using three bearings to find the observer's location.

moderate accuracy using the mile/kilometre recorder of a car, provided that it is possible to drive in a straight line between the points. Otherwise, the most accurate method is to use a tape, but a reasonable compromise is to pace out the distance. Before doing this you should have measured your average pace length by counting paces between two points at least 200 m apart, as measured by tape. Keep a note of your paces per hundred metres in your field notebook, and draw up a photograph scale on a piece of card, graduating it at intervals of 100 m or less.

It is also possible to calculate the scale by dividing the height *above ground* at which the photograph was taken by the focal length of the camera. Altitude above sea level and focal length are shown on the margin of some, but not all, aerial photographs. But even if these values are known it is also necessary to know the mean height of the ground above sea level. If you can estimate this figure from a map with contours or spot heights, or have the use of barometer altimeters, then:

$$\frac{\text{height of aeroplane above sea level} - \text{mean altitude of ground}}{\text{focal length of camera}} = \text{scale}$$

## 3.2   Finding location by compass bearing

The method is illustrated in Fig. 10e, and involves taking compass bearings on three well defined features that can be seen from the unknown location, and that are recognisable on the map or aerial photograph. By plotting these bearings as lines passing through the features on the map, an intersection is obtained which marks the position of the observer - but note the following:

(a)  The bearing must be taken carefully, read to the greatest accuracy possible (the nearest degree is not good enough), and checked by repetition.
(b)  Only two bearings will, of course, give an intersection, but a location obtained in this way is unreliable and should be checked with the third bearing. Often the three bearings do not give lines meeting exactly at a point, but make a small 'triangle of error'. The true location should lie within this triangle.
(c)  Because aerial photographs are likely to be distorted in various ways, it can happen that one obtains a large triangle of error however carefully the bearings are observed.

If three suitable features cannot be recognised, an alternative method is to take a single bearing on an identifiable feature not too far away and pace the distance to it along a straight line. For further advice on simple surveying methods, consult Compton (1962).

## 3.3    Strike and dip of planar structures

Measuring the dip and strike of a plane is one of the most basic methods available to a field geologist, and must be clearly understood. By these measurements one can define the attitudes of bedding, cleavage, foliation, banding, joints, the axial planes of folds and of any other structure which is a plane and not a line.

The strike of a plane may be defined as the direction (i.e. azimuth) of a horizontal line which lies within it (Fig. 13a). Strike is usually given a three-figure compass bearing between 000° and 180° (although there will be equivalent bearings between 180° and 360°).

The dip of a plane may be defined as the angle between the horizontal and the plane itself when measured at right angles to the strike direction (Fig. 13a). It is, in fact, the steepest direction within the plane - the direction in which a liquid would run if free to do so. Dip is recorded as an angular value together with a general indication of its downslope direction (N, S, NE, etc.). The need to give this general direction as well as the angular value may be understood by imagining a plane which strikes 060° (roughly northeastwards); if one says that this plane dips at 40° it is also necessary to state whether this is 40° towards the north-west or 40° towards the south-east, which are alternative possibilities. It is not necessary, however, to give the precise direction in which the plane dips since this will always be 90° from the strike direction. Thus the complete description of the attitude of inclined beds at a certain locality might read in the notebook as:

bedding:   strike 060° ; dip 40° SE.

Note that the planar structure measured (in this case bedding) must be specified. Symbols for showing the attitude of planar structures on the map are given in Appendix B.

There are several methods for measuring dip and strike: here we will describe how to do so in two situations. In the first, the planar structure to be measured is exposed as one or more flat surfaces excavated along its length (Fig. 11). In the second situation, the structure is either not exposed as flat surfaces or it is an imaginary plane as, for example, the axial plane of a fold (Fig. 12b).

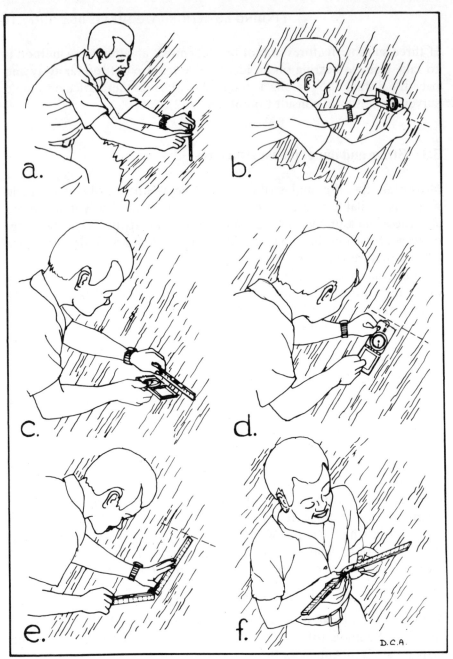

**Figure 11** Measuring strike and dip on an exposed surface of a planar structure: finding the horizontal direction using (a) a level; (b) a compass/clinometer - the clinometer is held in the vertical plane and rotated while in contact with the surface until the needle reads 0°; (c) reading the direction of strike with a compass and level; measuring the angle of dip using (d) a compass/clinometer; (e) a carpenter's folding rule; note that the protractor on the rule can be conveniently read after removal from the rock surface (f).

*Measuring the strike and dip of an exposed, flat surface.* The method is illustrated in Fig. 11. Select a surface which is both flat and appears to have an attitude typical of that locality. Make the measurement in the following stages:

(1)   Find the direction of a horizontal line on the surface with the help of a bubble level, or a clinometer set at 0° dip (Fig. 11a & b).

(2)   Measure the direction of this line and record it in the notebook (Fig. 11c). If you have used the dip needle in a Brunton or Silva compass to find a horizontal line, then you will need to mark this line on the surface before using the compass to measure its direction; scratching along the line with a fragment of rock will be enough.

(3)   Measure the angle of dip by holding the clinometer against the surface so that it makes a right angle with the horizontal line (Fig. 11d & e). Record this angle and the general direction of dip in the field notebook.

If the surface to be measured is uneven, an average orientation may be obtained by laying the mapping board flat against it and making the measurements on that.

Note that dip and strike may be measured on a overhanging flat surface, although handling the instruments is more difficult than on an upward-facing surface.

*Other methods of measuring strike and dip.* At many exposures it is easy to see the direction in which a planar structure is dipping, yet because there are no flat surfaces eroded along it the method described above is impossible (e.g. Fig. 12c). Exposures of this kind will show the trace of the planar structure as it crosses surfaces lying in two or more directions. On each of these surfaces the traces will display an 'apparent dip'; that is, will show an inclination which is less than the true dip (except in places where the surface is vertical and in the direction of dip). There are several methods for dealing with this situation:

(a)   The true dip can be calculated from measurements of the apparent dip on two surfaces which have different orientations (see Compton 1962, pp. 31, 32 and 362 for a full explanation). Although this method is accurate, we normally prefer one of the more straightforward procedures described below.

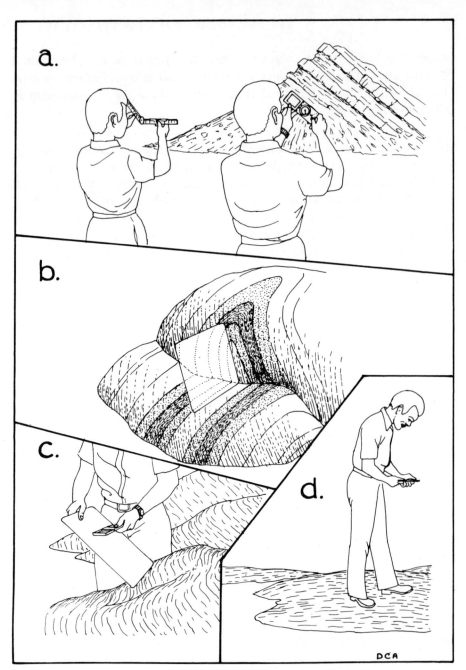

**Figure 12** Measuring strike and dip of a planar structure by sighting: (a) sighting the angle of dip using a folding rule or a compass/clinometer while looking along the strike direction; (b) using a mapping board to define and measure the axial plane of a fold; (c) using a mapping board to define the projection of a planar structure which is not exposed as flat surfaces; (d) sighting the strike of a planar structure exposed in a subhorizontal slab.

(b)  Hold the mapping board so that it lies in the same orientation as the planar structure by lining it up by eye with the apparent dips as seen on two or more surfaces of different orientation (Fig. 12c). Then measure the strike and dip of the board. This method is difficult without some assistance, but not impossible. It can also be used to measure the attitude of the axial plane of any fold which is exposed in such a way that the dips of the two limbs can be seen simultaneously. In this case the board is held in the position of the imaginary plane which separates the two limbs (Fig. 12b).

(c)  Search the exposure for a surface which is both horizontal and shows the trace of the planar structure. The direction of the trace on this surface will be the strike, which can be measured by sighting directly down onto it (Fig. 12d). To read the dip, stand back from the exposure and, looking along the strike direction, hold up the clinometer and sight the dip angle as seen from this direction (Fig. 11a).

(d)  As an extension of the last method, the direction of strike and angle of dip can often be sighted from a viewpoint which shows the planar structure on a large scale as it crosses a series of exposures or a hill. However, be sure you are looking directly along strike as you make the measurement, as an oblique viewpoint will give a false impression of the true dip.

Beginners in mapping often lack confidence in measurements of strike and dip which have not been made on an exposed surface of the planar structure. It is important to emphasise, however, that the various 'sighting' methods often provide a more accurate measure of local attitude of a structure than can be obtained from a single, small, flat surface, especially in areas where the planar structure is irregular or undulating.

It should also be said that on some nearly flat, slabby exposures it is possible to determine the direction of strike and see the general direction of dip without being able to measure it accurately. Such strikes and dip directions should be recorded, since they provide useful information in areas of few exposures.

## 3.4   Direction and plunge of linear structures

Common linear structures include the mineral growth lineations often seen as striations on the foliation planes of gneisses and schists, lineations produced by the intersection of bedding and cleavage in slates, the parallel corrugations on the cleavage of many phyllites, slickenside striations, and fold axes. The attitudes of such structures are described in terms of 'direction' and 'plunge'. The direction is the compass bearing of the structure as seen when viewed vertically from above, with the compass pointed in the direction in which the structure plunges downwards (Fig. 13b & c). Note that directions must be stated in terms of the whole 000° to 360° of the compass, unlike the strikes of planar structures, which can be given between 000° and 180°. The plunge of a linear structure is the angle between the structure and the horizontal as measured in the vertical plane (Fig. 13b). Like dip, plunge may be measured using a clinometer or carpenter's folding rule, remembering to hold the instrument vertically while making the reading (Fig. 13d & e). The measured attitude is recorded in the notebook in the form:

*crenulation lineation:*   direction 070°; plunge 50°.

The equivalent symbols to be placed on the map are shown in Appendix B.


## 3.5   Measurement of topographic height

In carrying out a geological survey, topographic heights are needed when describing geomorphological features, as well as for the construction of cross sections. Accurate topographic data are essential if any gravimetric work is to be done. If the base map lacks contours it will be necessary for the geologist to make his own estimates of heights. For some purposes, such as the description of features less than 50 m or so high, it may be sufficient to make a rough estimate by eye. This is best done by standing well back from the feature and using an object of known size (a tree, man or large boulder for example) which is on the feature itself as a scale. For larger features and more accurate work, two quick methods which are particularly useful are described below, and require either barometer altimeters or the making of simple triangulations. If a very high degree of accuracy is required, then a surveyor's level will be needed, but the description of the use of this instrument is beyond the scope of this book.

*Using barometer altimeters to measure height differences.* A barometer may be used to measure elevations because the difference in atmospheric pressure between two points is largely a function of their

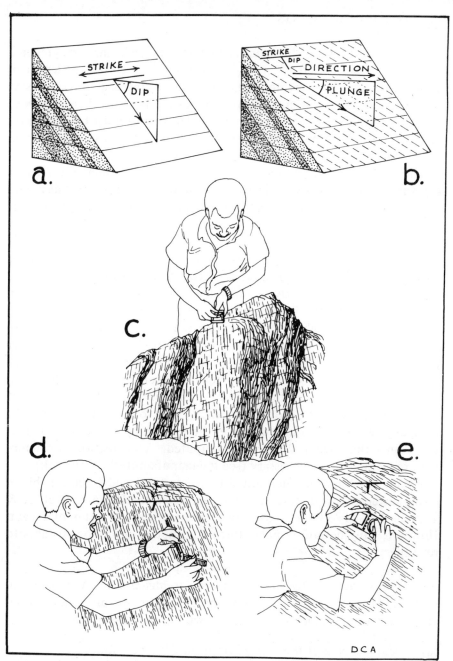

**Figure 13** Measuring the attitude of linear structures: (a), (b) the difference between the strike and dip of a plane and the direction and plunge of a linear structure; (c) using a compass to read the direction of a lineation - align the compass and lineation in the vertical plane; (d) measuring the angle of plunge using a folding rule with a level; (e) measuring the angle of plunge using a compass/clinometer - always hold the compass or the rule within the vertical plane.

difference in height. The aneroid barometers used in surveying are already calibrated to read in heights, often to a precision of less than 1 m. They can be used to measure height differences directly (say, between the base and top of a hill), but if absolute heights above ordnance datum (O.D.) are required, the barometers must be frequently calibrated against a point or points of known height (e.g. bench mark, triangulation point).

The main sources of inaccuracy in barometer estimates of height are variations in atmospheric pressure in the course of the day, and marked differences in temperature and humidity. Calm, stable weather provides the best conditions, especially in the early and late hours of the day.

Using only one instrument, the best results will be obtained by measuring the pressure difference between a known point and the unknown height as quickly as possible. To determine the height of an important station (the base camp, for instance) the comparison with a point of known height should be repeated several times. If two baro- meters are available, one of them is kept in the base camp (or another point of known height) and read at frequent intervals during the day by an assistant. The other instrument is taken out on a traverse to measure the pressure at the required points in the field. The time of each measurement is recorded in each case, so that it may later be compared with the reading made on the base camp instrument at that time, and the height difference computed. A correction for temp- erature is usually also necessary (see the manufacturer's handbook).

Finally, it should be mentioned that a method said to be capable of giving an accuracy of less than 1 m has been developed by Wallace & Tiernan Ltd (1967), and uses two base stations of different height. However, this method requires three barometers and two assistants to read them.

*Estimating height differences by simple triangulation.*  A very useful method of obtaining the approximate heights of prominent topo- graphical features depends on making two simple measurements of:

(a)   the distance $d$ of the feature from the observer; and
(b)   the angle $A$ subtended by the feature in the vertical plane, as seen from the point of observation.

The method is illustrated in Figure 14a, from which it is evident that the height $h$ of the feature is given by the formula:

$$h = d \tan A$$

The distance $d$ is easily obtained from the base map or aerial photograph. The angle $A$ subtended by the feature may be measured rather approximately by sighting along a clinometer, but for accurate measurements an optical clinometer (Fig. 1f), Abney level or similar instrument should be used. For good results the angle $A$ should not be too small, or the feature too distant. As an example, a hill 200 m high viewed from a distance of 1 km subtends an angle of over 11°, and an error as large as half a degree in measuring $A$ would give a height error of only about 9 m. The same hill viewed from 5 km subtends an angle of a little over 2°, and a half degree error in measuring it would result in over- or under-estimating the height by as much as 44 m. However, if these

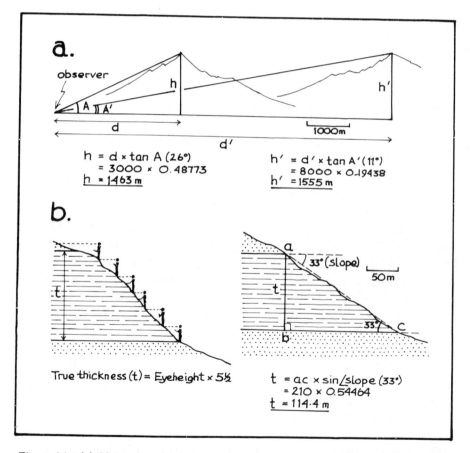

**Figure 14** (a) Measuring the heights of distant objects by triangulation. (b) Two methods of measuring the thickness of horizontal formations (or the height of small topographic features).

limitations are understood it is evident that even an estimate which is wrong by over 20% is likely to be more accurate than a naked eye estimate at a similar distance.

## 3.6   Measurement of true thickness

The true thicknesses of bedded formations and of other tabular bodies such as dykes, sills and veins should be determined whenever possible. The measurement of small units can be carried out with the help of a folding rule, or a staff of known length, or even with the handle of the geological hammer. Remember to make the measurement at right angles to the bedding planes or contacts. For larger bodies the method used depends partly on whether the beds are horizontal or have a significant dip.

*Measuring thicknesses on horizontal or subhorizontal beds.*   If the beds to be measured are horizontal or dip at only a few degrees, one can use any method capable of determining the height of the outcrop. One straightforward but not very accurate method is to employ the eye height of the observer in the way illustrated in Figure 14b. Eye height should have previously been measured by marking on a wall or tree the height at which one's eyes naturally come to rest when looking straight ahead - it is a good idea to keep a note of both eye height and pace length at the front of the field notebook.

   If the exposure is steep enough, the vertical height can be measured by lowering a measuring tape or rope from the upper contact. Otherwise the best method is to tape or pace out the distance down slope between upper and lower contacts (ac in Fig. 14b) and use a clinometer to estimate the average slope between these two points. This will allow computation of the true thickness (*t*) by simple trigonometry. A worked example is shown in Fig. 14b.

*Measuring thicknesses on dipping beds.*   One simple way to measure true thicknesses on an outcrop of dipping rocks is to use a clinometer attached to a staff at about eye height, and set at the angle of dip - the method is illustrated in Figure 15a. Although it is rather awkward in practice, this method has the advantage of giving a direct answer, without the need of corrections for dip.

   Other ways of tackling the problem are shown in Figures 15b-d: they require the measurement of dip angle, slope angle and the distance (ac) between the upper and lower contacts of the unit. Alternatively the

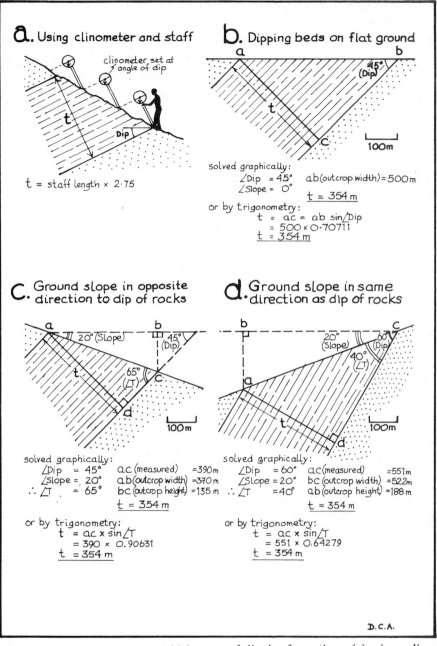

**Figure 15** Measuring the true thicknesses of dipping formations: (a) using a clino-meter and staff; (b) by triangulation.

vertical separation of the upper and lower contacts can be measured using the eye height method (i.e. bc in Fig. 15c and ab in Fig. 15d), or the outcrop width may be read from the map. However, beware of using aerial photographs to determine outcrop width, since they are liable to severe distortion if slopes are steep. Having obtained these measurements, the true thickness can be determined either graphically or by simple trigonometry. Worked examples of several common situations are given in Figures 14 and 15.

Estimation of the thickness of very large units is often best done on vertical sections drawn along the direction of dip, and using measured topographic heights to make the profile along the section line accurate. Employ the same vertical and horizontal scales so that the angles need not be changed. It is usually possible to use for this purpose the sections drawn for structural interpretation of the area.

# 4 The final map and report

## 4.1 Production of the final map

The results of a field mapping project are usually presented as a re-drawn version of the field map and an explanatory written report. The original field maps should also be preserved for reference. The final map differs from the field map in that, in the course of redrawing, the details recorded in the field are edited, simplified and combined together to make the map clearer and more easily understood. Exposure numbers and many of the original lithological notes may be omitted, but as much structural information as possible should be recorded unchanged. The outlining of individual exposures may or may not be retained on the final map, according to personal taste, but in any case the degree of certainty of boundaries, as expressed by the variation from solid to broken lines, should be carefully preserved.

*General method.* We assume here that the fieldwork has been recorded on sections of a base map (Section 2.3), which can be assembled into a single sheet for copying. If, on the other hand, the field observations are still on separate aerial photograph overlays, then a topographic base must first be prepared by making a photo-mosaic, taking care to keep observed north arrows parallel and using any surveyed fixed points as controls.

Before starting to draw the final map, a number of decision have to be taken on aspects which affect the way in which it is to be produced:

(a)  Is the final map to be at the *same scale* as the field map or *reduced*? Reduction is best done photographically or by an optical reducing machine after transferring the data from the field sheets, but before adding any colouring. Remember to keep all numbers, lettering, symbols and ornaments large enough for them to be legible after reduction.

(b)  Consider whether one copy of the map or several are required. If only one copy is needed, the final map may be made by direct tracing on a sheet of strong paper (assuming the availability of a tracing table), but if several copies are needed, draw the first on tracing paper, polyester sheet or tracing linen so that it can be reproduced by the dyeline process.

(c)   Is the main map to include all the mapped information, or are there to be separate maps for, say, specimen localities, hydrogeology, topography or structure? Note that while a separate structural map can be very useful in a complex area, this does not mean that basic structural information can be omitted from the main map.

(d)   Carefully design the *layout* of the main map remembering that, as well as the map itself, space will be needed for a key, location map, cross section(s), title, and a wide border (say, 3 cm or more). If possible, the map should be oriented with north towards the top of the sheet, but if the map is an awkward shape it may be necessary to use some other orientation.

(e)   Consider how much *topographical detail* is to be included on the geological map. As a guide, one should show enough topography to allow another geologist to locate on the ground all the geological features shown, but not so much as to obscure the geology itself. It is usually useful to show major roads, important tracks, wells, villages, the main watercourses, hill summits, and the main break of slope around the foot of hill masses. If the area also possesses interesting geomorphological features which you wish to record, then a separate geomorphology map is the best solution.

Having made these decisions, trace off the required details from the field map. Topography, boundaries and symbols can be traced directly in ink, but take great care to maintain the accurate orientation of the structural symbols. If there is any doubt about this it is best to re-plot the symbols using a protractor. Numerals and letters are transferred initially in pencil, and only inked in when it is clear that they are accurate and in the best position to be easily read.

The next stage is to ornament or colour in the various formations, and add a key, reference grid, scale, north arrow, cross section(s), and location map. Each of these additions requires careful consideration.

*Ornament or colours.* Each formation should be clearly differentiated from others by a distinctive ornament or coloration. Whether it is better to use ornament or colour depends on the number of copies to be made and the method of reproduction. Undoubtedly a coloured map is more attractive and easier to understand than one which has only black and white ornament, and if only a small number of copies are needed, or a colour-printing press is available, then this is the better method to use. To colour in the outcrops, coloured pencils are much easier to use than watercolours, which will cause the paper to wrinkle unless it has been dampened and stretched beforehand. When applying colours keep them pale in tone, since deep shades obscure the geology and topography, and reserve the strongest colours for those formations which have the smallest outcrops. Apart from this, one should follow the usual convention whereby, for instance, limestones are usually represented by shades of blue, sandstones by yellow or brown, acidic igneous intrusions by orange or red, and so on. Precise identification may be obtained by adding numbers or letters on top of the colours. In some schemes, the same colour is used for more than one formation, and these are then distinguished from each other by adding overall ornaments. Ideas on black and white ornament may be obtained from maps published in journals. They may be drawn in pen, or cut from self-adhesive sheets of printed ornament - the latter are neat, but easily peel off if handled extensively. In any case, maps carrying such ornament should never be folded.

*Map key.*   Prepare a separate pencil sketch of the key before adding it to the map. It must include all the formations or rock types shown separately on the map, usually by listing their ornaments in a series of small boxes against each of which is the appropriate name (see Fig. 16b). The list should be in stratigraphical order, with the youngest rocks at the top, although intrusive igneous rocks are often shown separately from sediments and volcanic rocks. Every symbol used on the map must also appear in the key, with its explanation. It is usual to list the symbols below the list of rock types. Note that it is all too easy to omit a formation or symbol from the list, so check carefully before adding the key to the map.

*Reference grid.*   When writing the report, it will often be necessary to refer to specific localities on the map. It would take the reader far too long to find individual locality numbers, and in any case these are best omitted from the map for the sake of clarity. It is better to use a grid system to which localities can be referred numerically - a simple example is shown in Figure 16a. Some countries have a national grid system which can be used for this purpose. Alternatively, if accurate base maps are available, it may be possible to use latitude and longitude as the grid. If this is not possible the geologist must draw a grid of his own to cover the area mapped. Where possible, however, some points on the map should be related to the latitude and longitude system, so that when, at some later date, good topographical maps become available the geology can be accurately transferred onto them.

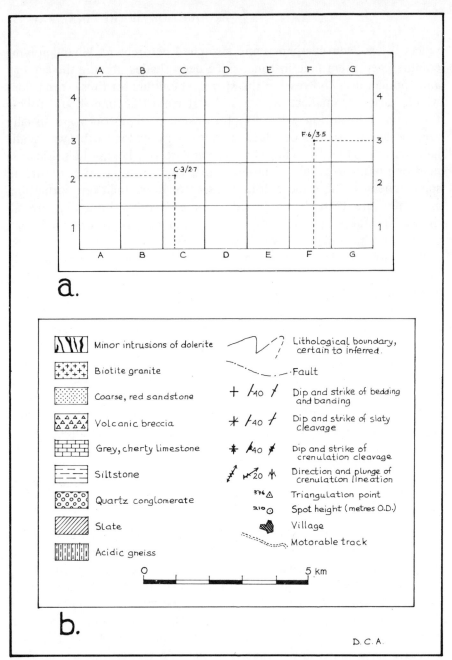

**a.**

**b.**

Minor intrusions of dolerite

Biotite granite

Coarse, red sandstone

Volcanic breccia

Grey, cherty limestone

Siltstone

Quartz conglomerate

Slate

Acidic gneiss

Lithological boundary, certain to inferred.

Fault

Dip and strike of bedding and banding

Dip and strike of slaty cleavage

Dip and strike of crenulation cleavage

Direction and plunge of crenulation lineation

Triangulation point

Spot height (metres O.D.)

Village

Motorable track

0           5 km

D. C. A.

**Figure 16** (a) A grid reference system of location, showing the references of two points as examples. (b) Design for a simple key for a geological map.

*Scale.* Every map should have a scale bar shown prominently upon it (see Fig. 16b). The scale may also be stated numerically (e.g. 1 : 20000), but remember, if the map is to be reduced then this value must correspond to the scale of the final product. The scale bar should always be calibrated in the metric system, but miles and yards may also be shown in regions where they are still in common use.

*North arrow.* Like scale, the direction of north should always be indicated clearly on the map. Show either true north alone, or true north together with magnetic north for a stated year. The advantages of orientating the map such that north is towards the top have already been mentioned, as when this is done the reference grid can be drawn with north as one of the grid directions.

*Cross section(s).* One or more cross sections drawn in the vertical plane are essential to the presentation of most geological maps. They may be drawn on the same sheet as the map itself, or shown separately in the report. If drawn with the map they will normally be given the same horizontal scale. A moderate amount of vertical scale exaggeration may be used if necessary, but true vertical scale is better where possible. Remember that exaggeration of the vertical scale will also distort all dips. Cross sections contain a greater element of interpretation than does the map itself. It is often useful to extend this interpretation some distance 'up into the air' as well as 'down into the ground'; that is, to suggest the former upward continuation of the structures above the present level of erosion by means of dotted continuations of the boundaries.

If the area contains a complex stratigraphical succession, it may be a good idea to include on the map a vertical strip section of the succession, drawn on an exaggerated scale and showing details of the lithology and fossil content.

*Location map.* The location map shows the general position of the mapped area and its boundaries on a small-scale map of the region or country in which it lies. This map may be placed in one corner of the main geological map or given separately in the early part of the report.

1 *Determine first the scale of the finished map and how many copies are needed.*
2 *Before starting to draft the final map, make a detailed design of its layout.*
3 *Always include a key, reference grid, scale, north arrow and location map. Cross sections also may be included, or given separately within the report.*

## 4.2    Writing the geological report

The report is partly an explanatory description of the geological map, expanding upon, and giving further details of, the various features it shows, as well as including any extra information which may have been obtained from laboratory examination of collected material. It should also provide a general synthesis of the geology, and suggest explanations for the various geological relationships described. The report is, however, dependent upon the map, in that it cannot be properly understood without it, whereas the map should be drawn so as to stand on its own.

*Planning.* Although the structure and length of the reports vary according to the purpose of the projects described, the layouts of that kind of report which covers the main geological features of a newly mapped area tend to follow a common pattern, as can be appreciated by examining the reports of national geological surveys. A typical layout may contain the following items:

Title page
Contents
List of illustrations
Abstract
Section 1    Introduction
Section 2    General geology and geological setting
Section 3    Field relations and lithology
Section 4    Structural geology
Section 5    Palaeontology
Section 6    Economic potential
Section 7    Geological history
Section 8    Summary and conclusions
References

Each of the sections (1-8) will usually be divided into several titled subsections, and these may be further divided. The layout may, of course, be varied according to need and personal preference, but the sections must follow each other in some kind of logical order. It is important to plan the layout in detail before starting work, to determine not only the various chapter headings but also how these are to be subdivided. Rough estimates should be made of the length of each section and subsection. Reports should not be made any longer than necessary, and it is therefore best to write each part with some length limitation in mind. A reasonable estimate of the likely word- and page-length is that one side of A4 paper (30 cm x 21 cm) will carry about 300 words when typed double-spaced. For most student mapping projects, a report length of about 5000 words (i.e. about 17 pages of typed A4) is reasonable. The sheet-map reports of professional survey geologists are naturally much longer, reaching 25000-30000 words or more. To estimate the lengths of individual sections, therefore, take the figure for the maximum total number of words and divide it between the various chapters and sections according to what you believe to be the relative importance of these topics in the area mapped. Thus, in one area, the economic potential may be considerable and deserve a long chapter, whereas in another the economic possibilities may be limited but the structure very complex, requiring a different balance between the chapters. While making a plan of the layout, subjects which need to be illustrated should also be listed, as should the titles of any tables which are to be inserted into the text. Although this preliminary plan will no doubt need later modification, it is likely to prove a useful guide during the writing of the report.

*Style.* Everyone can develop his own writing style, although perhaps the only way to learn is by trial and error, and by following good examples. It is a difficult subject to teach, but there are several points which we believe should be emphasised here. First of all, bear in mind the readers to whom the report is addressed. Generally these will be fellow geologists and so you can assume a general knowledge of terminology and ideas, but do not assume knowledge of the local, or even regional, geological setting. Secondly, the writing should be concise and accurate, but expressed in an interesting way so as to retain the reader's attention. Long, descriptive passages can easily become boring, so vary them by using tabulations to record some of the facts, and diagrams to display the data in easily understood form. These methods also save space. Where appropriate, quantify descriptions (of, for example, grain size or proportions of constituents) but not so much as to overwhelm the reader with numbers. Be accurate in spelling (use a dictionary) and avoid frequent repetition of words or phrases - Roget's *Thesaurus* provides useful and accessible alternative words. It is also essential to distinguish clearly between the facts and the interpretations placed upon them; for example, complete the description of a feature before discussing its possible origin. Indeed, to a great extent there should be a concentration of the factual material into the earlier part of the report and of interpretation into the later sections.

Finally, do not become discouraged by the effort needed to attain a good style. Few people find writing an easy task, and almost everyone has to make at least one preliminary draft before being able to write the final version.

*Title page.*   The title should be as informative and as short as possible. It should state the general nature of the project and the geographical name of the area and region in which the fieldwork was carried out. Below the title should appear the author's name, the name of the institution to which he belongs, and the year.

*Contents and list of illustrations.*   These are compiled from the final version of the manuscript and include all heading and subheading titles, followed by a list of the captions of all figures, diagrams, plates and tables. Against each item should appear the corresponding page number.

*Abstract.*   The abstract is also normally written after the main text has been completed. It must be kept short (100-600 words, depending on the length of the report) and it consists of a highly condensed account of the objectives and chief results of the project. Lay particular emphasis on any new or especially significant evidence, and summarise any new hypotheses proposed. Remember that the abstract will often be read on its own, or in conjunction with the final summary and conclusions.

*Introduction.*   This section is often best begun by giving the precise location and extent of the area studied (quote latitudes and longitudes, and refer to the location map), with comments on its accessibility. A clear statement of the aims of the investigation should then follow, as should mention of the time spent on the project and the methods used to attain the stated objectives. It is also useful to include here a brief summary of previous work, quoting authors and dates for the key references but reserving their detailed results for comment in the main text. Any general remarks on topography, climate, vegetation and culture of the area may also find a place here, although any systematic account of the geomorphology should be reserved for the following chapter. It is customary to conclude the introduction by formally acknowledging any academic or personal help given to the writer by individuals or institutions.

*The main text.*   The main body of description and interpretation which follows the introduction will vary so greatly in content from project to project that it would be difficult to offer detailed advice. We therefore confine our remarks to a number of special points.

The main text usually begins with a description of the general geology, and under this heading it is necessary to include some account of the regional geology, so that features of the area mapped can be seen in relation to those of the surrounding country. Moreover, in order to appreciate the geology in its temporal setting it is useful to give a summary of the stratigraphical succession, even though this depends largely on the interpretation of evidence to be given later. One method of doing this is to include at this point a table that summarises the names, general characteristics and interrelationships of the various formations mapped. This tabulation may also indicate the sequence of events which the succession records, and suggest correlations with adjacent areas. A more detailed discussion of the geological history can be given near the end of the report, where the rocks can be discussed in terms of sequences of sedimentary environments, magmatic events, metamorphism and structural evolution. The evidence described in the preceding chapters can be drawn upon to support this interpretation of the geological history, but any uncertainties must also be pointed out. Indeed, any good scientific work will raise new problems, even as it solves some of the old ones, and these are better emphasised than passed over.

In the discussion of economic potential, all known occurrences of useful minerals and rocks should be described in full, not forgetting to quote in detail from any previous work on this aspect of the area. Water supplies should also be considered, particularly in arid areas, as should the presence of any features which might give rise to geological hazards. If you wish to encourage further investigation of some economically interesting mineral occurrence, look at the situation from the point of view of a potential developer and assess the problems of access, water supply, local labour and the like.

The final chapter of the main text is primarily a synthesis of the results and main conclusions of the work. Such a synthesis may seem unnecessary to the author of the report, who will by now be all too familiar with the details of the project, but to the reader this section can be essential in drawing the report to an understandable and logical conclusion. Moreover, at this point the geology of the mapped area may once again be placed into its setting, noting ways in which the new work may help to answer questions on a regional scale. Finally, it is useful to suggest ideas for future work in the area.

*Illustrations.*   Carefully prepared illustrations are a very effective aid to the clarity and attractive presentation of a report, but poor illustrations detract from it even more obviously than a weak text. Illustrations are particularly good at showing local field relations, petrography, fossils, complicated structures, and for visualising the relationships between large amounts of numerical data. Many types of illustration may be used, including redrawn field sketches and sketch maps, photographs of field relations and of material in the laboratory, graphical logs of stratigraphical successions, block diagrams, cross sections, and plots of chemical, structural or textural data in the form of binary and triangular graphs, histrograms and stereograms. Select the most significant and clear material. Note that only the very best field photographs are better than a good field sketch (or even than a sketch traced from the photograph itself). Thin section drawings, or tracings, are much clearer than most microphotographs. Consider also the method of reproduction (xerox, dyeline, printing press etc.) and whether the use of colour would improve your diagrams. On the other hand, simple line drawings with no unnecessary detail can often be the most effective. Finally, we suggest that you:

(a)  Keep illustrations large enough for details and lettering to be easily seen.
(b)  Make preliminary sketches for the layouts of complicated diagrams.
(c)  Remember to include a scale or show orientation wherever appropriate.
(d)  Fully label any features which are not explained in a key or in the caption.

*References to literature.*   In writing the report every effort should be made to refer to relevant literature, especially of course when discussing previous work on the area. Moreover, any description, list of data or hypothesis which is not the author's own should be acknowledged by citing its source.

When citing an author in the text both his name and the year of publication should be given, using brackets in the following way:

. . .and it is reported that similar deposits occur along the same fault to the north-east (Yusif 1972)

. . .and according to Yusif (1972), similar deposits occur along the same fault to the north-east

Note also that if the publication cited has two co-authors then both are given in the text:

. . .these intrusions are said to be younger than the Nubian sediments of the area (Awadalla & Smith 1980)

whereas if there are more than two co-authors the in-text reference is given as:

. . .according to Louchon *et al.* (1969)

with the full list of authors given in the reference list.

All citations given in the text are quoted in full in the References, placed at the end of the report. Such lists are usually arranged in alphabetical order by authors' names (the 'Harvard system), but various styles are in use, as can be seen by inspecting a selection of survey reports, journals and textbooks. One common system is illustrated by the two examples below, of which the first is the full reference to a paper published in a journal and the second refers to a book:

Frey, F. A. 1969. Rare earths in a high temperature peridotite intrusion. *Geochim. Cosmochim. Acta* 33, 1429-447.
Wyllie, P. J. 1971. *The Dynamic Earth.* New York: Wiley.

That is, the reference to a journal is given in the following order:

Author's name, followed by the initials of his given names, year of publication, title in full, abbreviated name of the journal (in this case *Geochimica et Cosmochimica Acta*), volume number, first and last page numbers of the article. The title is commonly either underlined or printed' in italics, and the volume number either underlined or printed in bold type (depending on the method of reproduction).

Some difficulty is caused by the different conventions used in abbreviating journal titles. The most generally accepted system is that given in the *World list of scientific periodicals* (1963), but this three-volume work is no longer commonly available, and reference can be made instead to *Short titles of commonly cited scientific journals,* published by the Royal Society. Note that when referring to books it is the title that is underlined or printed in italics, and this is followed by the location and name of the publisher.

As an alternative to a 'list of references' one may include a 'bibliography', as at the end of this book. This differs from a list of references in that it is selected on the basis of relevance to the subject under discussion, and not all the references given in it will be cited in the text.

The advantage of a bibliography is that it can give a more complete and balanced picture of the background material available. Listing only references cited is, however, generally regarded as more satisfactory for mapping reports and most scientific papers.

1   *Plan the report before starting to write, defining all headings, estimating the length of each section, and deciding the approximate number and contents of figures, tables and photographs.*
2   *Distinguish between facts and the interpretations placed upon them.*
3   *Aim at a clear and concise style of writing.*
4   *Use tables and figures in ways which both clarify the text and save unnecessary words.*
5   *Emphasise, rather than cover up, any special problems or points of uncertainty in the geology.*
6   *Acknowledge previous work and ideas, and take great care in compiling a full and accurate list of references.*
7   *Ruthlessly revise the first draft.*

# Appendix A  Field equipment and supplies

The following lists can only be general guides and should be modified to suit local conditions and personal preferences.

## A.1  Mapping equipment

compass
clinometer
altimeter(s)
hammers
chisels
safety goggles
hand-lens
hard-backed field notebook
hard pencils
pens
pencil sharpener/razor blade/knife
camera, films, flash attachment
leather case for carrying pencils, note-
    book, etc.
mapping boards/mapping case

base maps
aerial photographs, photo-mosaics,
    satellite photo prints
pocket stereoscope
measuring tape (about 25 m)
water bottles
small bag or rucksack (strong enough
    for carrying rock specimens)
felt-tip marking-pens and sticky tape for
    labelling specimens
specimen bags or newspaper for
    wrapping
small bottle of dilute hydrochloric acid
    (tightly stoppered, in a plastic bag)
small magnet

## A.2  Base camp office equipment

mapping pens
lead pencils (various hardnesses)
coloured pencils
rubber eraser
rule/scale
protractor
reference books

waterproof inks (black and coloured)
tracing paper/tracing film
Sellotape
drafting tape
lined and plain writing paper
electronic calculator and spare batteries
bench stereoscope

## A.3  Camp equipment

tents (with poles, guy ropes and pegs)
groundsheets
folding beds
folding chairs
folding tables
food boxes
water containers

pressure lamps/lanterns/generator
electric torches and spare batteries
bedding
mosquito nets
spade, hoe, axe
buckets, dishes
long rope for drawing water from wells

## A.4  Vehicle equipment

containers for fuel (drums, jerricans)
oils for engine, gearbox, differentials
plastic hoses for siphoning fuel and
    water
brake fluid
grease
battery acid and distilled water
lifting jacks
tool-kit with spanners, etc.
valve tool and tyre pressure gauge
foot pump
assorted nuts, bolts, split pins, nails,
    wire
shovel
sand trays

towing rod or chain
ropes for lashing on loads
coolers for drinking water
spare parts (these might include: fan
    belt, distributor points, spark plugs,
    hoses and hose clips, mainspring
    leaves, spring shackle bolts, rear
    half-shaft, engine mounting brackets,
    tyres and inner tubes, hot vulcanising
    patches and clamp)
soap (normally carried for other
    purposes) can be mixed with cotton
    wool to temporarily seal leaks in
    petrol tanks or radiators

## A.5  Food and cooking equipment

Make some allowance for unforeseen delays. Dried foods are small in bulk and
weight, and are long-lasting. A basic list might include the following:

flour
baking powder
sugar
tea
coffee
salt
pepper
spices
rice
macaroni
lentils
beans

peas
cooking oil
onions
dried meat and fish
tomato paste
dried milk
custard powder
salted cheese
dried fruits
groundnuts
toasted bread

Plus fresh food as available: take plenty of plastic bags.

pans and cooking pots

cups

plates

dishes

kettle

knives, forks, spoons

tin opener

kerosene/petrol/charcoal stove,
    with fuel

matches

washing soap and soap powder

water-purifying tablets

insect spray

dishcloth

drying cloth

## A.6  First aid and medicines

first aid guide

antiseptic liquid or cream

Elastoplast tape and dressings

aspirin

codeine

antimalarial drugs

bandages (strips, triangles, elastic)

cotton wool

safety pins

kaolin mixture

senna pods

vitamin tablets

For other medicines ask a doctor for advice. It is unwise to take antibiotics, anti-histamines and the like indiscriminantly. Serums for treating snake bites and scorpion stings are available, but need to be kept for most of the time in a refrigerator.

## A.7  Personal effects

field clothes, including strong boots

sleeping bag

torch and spare batteries

pocket knife

sun spectacles

needles and thread, spare buttons

towel

toilet articles

money

reading books

transistor radio

# Appendix B    Map symbols and abbreviations

## B.1  Map symbols

There is no internationally accepted system of symbols for geological maps but those listed in Figures B1 and B2 will be widely understood. In any case, *all* symbols used on the map should also be shown and explained in the key. In order to cater for different needs, we have divided our list of symbols into two parts. Thus Figure B1 contains sufficient basic symbols for most student mapping projects and for other mapping projects on which, for reasons of scale, time or emphasis, there is no intention to record detailed structural information. Figure B2 gives a more comprehensive cover of these specialised symbols and is particularly intended for use in the detailed mapping of highly deformed rocks. In compiling this second list we owe a particular debt to discussions with Dr M. J. Fleuty of the Polytechnic of North Staffordshire, England.

Note also that it is often useful to distinguish different categories of symbol by using coloured inks; for example, black for lithological symbols and abbreviations, red for structures, and blue for topography.

## B.2  Abbreviations

We list below some suggested abbreviations which can be used on field maps and in field notebooks - many of these are in common use. Avoid making abbreviations which are too obscure, since field maps and notebooks should be comprehensible to other geologists.

*Rocks and minerals*

| | | | |
|---|---|---|---|
| agglomerate | aggl. | granite | gr. |
| alkali feldspar | alk. feld. | hornblende | hb. |
| andalusite | andal. | hornfels | hf. |
| andesite | and. | ignimbrite | ig. |
| basalt | bas. | kyanite | ky. |
| biotite | bi. | limestone | lst. |
| breccia | brc. | marble | mbl. |
| conglomerate | congl. | microgranite | mgr. |
| cordierite | cord. | migmatite | migt. |
| dolerite | dol. | mudstone | mdst. |
| gabbro | gab. | muscovite | musc. |
| garnet | gar. | mylonite | myl. |
| gneiss | gn. | pegmatite | peg. |

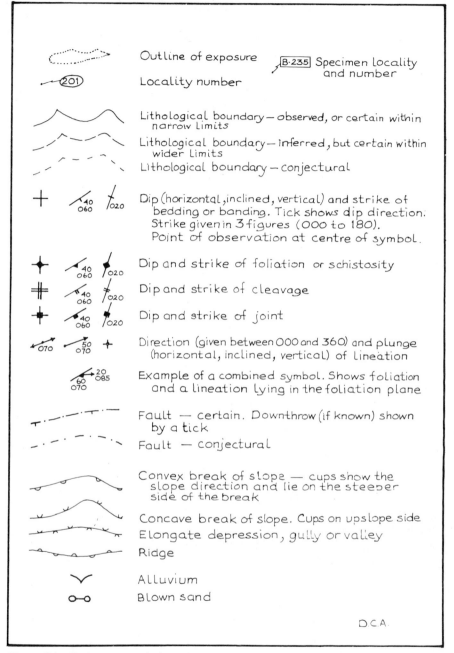

Outline of exposure

B-235 Specimen locality and number

Locality number

Lithological boundary – observed, or certain within narrow limits

Lithological boundary – inferred, but certain within wider limits

Lithological boundary – conjectural

Dip (horizontal, inclined, vertical) and strike of bedding or banding. Tick shows dip direction. Strike given in 3 figures (000 to 180). Point of observation at centre of symbol.

Dip and strike of foliation or schistosity

Dip and strike of cleavage

Dip and strike of joint

Direction (given between 000 and 360) and plunge (horizontal, inclined, vertical) of lineation

Example of a combined symbol. Shows foliation and a lineation lying in the foliation plane

Fault — certain. Downthrow (if known) shown by a tick

Fault — conjectural

Convex break of slope — cups show the slope direction and lie on the steeper side of the break

Concave break of slope. Cups on upslope side

Elongate depression, gully or valley

Ridge

Alluvium

Blown sand

D.C.A.

**Figure B1**   Basic map symbols.

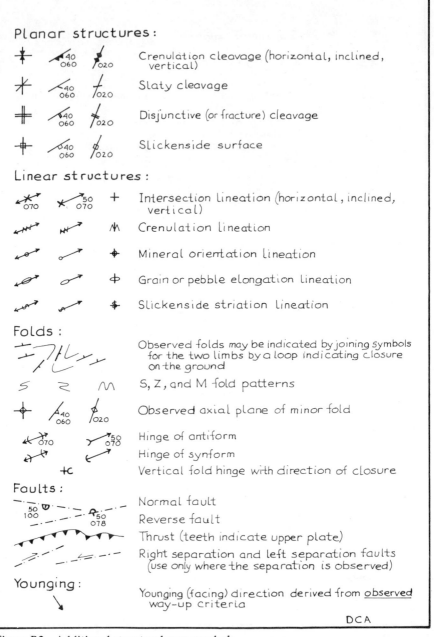

## Planar structures:

| | | | |
|---|---|---|---|
| ✚ | 40/060 | /020 | Crenulation cleavage (horizontal, inclined, vertical) |
| ✳ | 40/060 | /020 | Slaty cleavage |
| ╫ | 40/060 | /020 | Disjunctive (or fracture) cleavage |
| ✛ | 40/060 | /020 | Slickenside surface |

## Linear structures:

| | | | |
|---|---|---|---|
| 070 | 50/070 | + | Intersection lineation (horizontal, inclined, vertical) |
| | | M | Crenulation lineation |
| | | ✦ | Mineral orientation lineation |
| | | Φ | Grain or pebble elongation lineation |
| | | ✦ | Slickenside striation lineation |

## Folds:

| | | |
|---|---|---|
| | | Observed folds may be indicated by joining symbols for the two limbs by a loop indicating closure on the ground |
| S | Z | M | S, Z, and M fold patterns |
| ✛ | 40/060 | /020 | Observed axial plane of minor fold |
| 070 | 50/070 | | Hinge of antiform |
| | | | Hinge of synform |
| +c | | | Vertical fold hinge with direction of closure |

## Faults:

Normal fault

Reverse fault

Thrust (teeth indicate upper plate)

Right separation and left separation faults (use only where the separation is observed)

## Younging:

Younging (facing) direction derived from <u>observed</u> way-up criteria

DCA

**Figure B2**  Additional structural map symbols.

| | | | |
|---|---|---|---|
| phyllite | phyll. | sandstone | sst. |
| plagioclase | plag. | schist | sch. |
| pyroxene | px. | sillimanite | sill. |
| quartz | qtz | siltstone | stst. |
| quartzite | qtzite | slate | slt. |
| rhyolite | rhy. | syenite | sy. |

*Descriptive terms*

| | | | |
|---|---|---|---|
| amygdaloidal | amyg. | folded | fldd |
| arenaceous | aren. | foliated | flotd |
| argillaceous | argil. | jointed | jntd |
| bedded | bdd | linear | linr |
| cross bedded | xbdd | lineated | lintd |
| interbedded | ibdd | planar | plnr |
| calcareous | calc. | porphyritic | porph. |
| cleaved | clvd | veined | vnd. |
| crystalline | xtline | vesicular | vesic. |
| | | volcanic | volc. |

*Others*

| | | | |
|---|---|---|---|
| about | *c.* | occasional | occ. |
| approximately | approx. | parallel | // |
| compare | cf. | such as | viz. |
| especially | esp. | that is | i.e. |
| for example | e.g. | therefore | ∴ |
| locality | loc. | | |

# Selected bibliography

Allum, J. A. E. 1966. *Photogeology and regional mapping.* Oxford: Pergamon. This short text forms a useful introduction to interpretative techniques.

Barnes, J. W. 1981. *Basic geological mapping.* Milton Keynes: Open University Press.

Briggs, D. 1977. *Sediments.* London: Butterworths. Intended mainly as a handbook for project work on sediments in relation to geomorphology, this is also a useful introduction for geologists mapping clastic sedimentary rocks.

Compton, R. R. 1962. *Manual of field geology.* New York: Wiley.

Fleuty, M. J. 1964. The description of folds. *Proc. Geol Assoc.* **75**, 461-92.

Geological Society of London 1972. A concise guide to stratigraphical procedure. *J. Geol Soc.* **138**, 295-305.

Hobbs, B. E., W. D. Means and P. F. Williams 1976. *An outline of structural geology.* New York: Wiley.

Lattman, L. H. and R. G. Ray, 1965. *Aerial photographs in field geology.* New York: Holt, Rinehart & Winston. This short text concentrates more on manipulation of aerial photographs than on interpretation and so is complementary to Allum (1966).

Moseley, F. 1981. *Methods in field geology.* San Francisco: W. H. Freeman. An interesting discussion of the philosophy and general approach to mapping under a variety of conditions.

Peters, W. C. 1978. *Exploration and mining geology.* New York: Wiley. A guide to the methods of ore geology, with plenty of practical advice.

Read, H. H. 1970. *Rutley's elements of mineralogy,* 26th ed. London: George Allen & Unwin. Useful for the field identification of minerals.

Roget, P. M. 1970. *Thesaurus of English words and phrases.* London: Penguin. A valuable aid in finding the right word when writing reports.

Royal Society 1980. *Short titles of commonly cited scientific journals.* London: Royal Society. This booklet lists the abbreviated forms of many journal titles, following the *World lists* abbreviations system.

Tucker, M. 1982. *The field description of sedimentary rocks.* Milton Keynes: Open University Press.

Wallace & Tiernan Ltd 1967. *Precise altimetry.* Technical Publication B.P. 610.400. Some ideas for increasing the usefulness of barometric altimeters in field surveys.

*World list of scientific periodicals,* 4th ed, 3 vols 1963. London: Butterworths. The abbreviation system used in this list is the most generally accepted standard for scientific papers, but this work is not commonly available.